Adults, Mathematics and Work

Adults, Mathematics and Work

From Research into Practice

By

John J. Keogh, Theresa Maguire and
John O'Donoghue

BRILL
SENSE

LEIDEN | BOSTON

All chapters in this book have undergone peer review.

Library of Congress Cataloging-in-Publication Data

Names: Keogh, John J., author. | Maguire, Theresa, author. | O'Donoghue, John, 1946- author.
Title: Adults, mathematics, and work : from research into practice / by John J. Keogh, Theresa Maguire, and John O'Donoghue.
Description: Leiden ; Boston : Brill, Sense, [2018] | Includes bibliographical references.
Identifiers: LCCN 2018032007 (print) | LCCN 2018036664 (ebook) | ISBN 9789004381766 (ebook) | ISBN 9789004381742 (pbk. : alk. paper) | ISBN 9789004381759 (hardback : alk. paper)
Subjects: LCSH: Mathematics--Social aspects. | Mathematics--Psychological aspects.
Classification: LCC QA10.7 (ebook) | LCC QA10.7 .K46 2018 (print) | DDC 510.1/9--dc23
LC record available at https://lccn.loc.gov/2018032007

ISBN 978-90-04-38174-2 (paperback)
ISBN 978-90-04-38175-9 (hardback)
ISBN 978-90-04-38176-6 (e-book)

This book is printed on acid-free paper and produced in a sustainable manner.

For adult learners everywhere;
May you recognise and embrace the power of your knowing

CONTENTS

DIANA COBEN

FOREWORD

Haiku: 'Invisible mathematics'

Mathematics – abstract,
Powerful – yet in daily
Use often unseen

Mathematics often goes unrecognised in use – as if Harry Potter's invisibility cloak wraps around the mathematics we do when we're doing something else, so that we see only the 'something else' and not the mathematics involved. This book explores this paradox, focusing on adults using mathematics in the workplace. On the basis of this exploration it offers a Workplace Context Complexity Protocol: a tool for the analysis of mathematics in use in workplaces, calibrated according to its contextual complexity. The Protocol can be applied in a number of ways to identify what a person knows by experience and enables them to use mathematics effectively in their work. It enables workers and their managers to recognise and address the mathematics required and exercised in work tasks and discusses the implications for learners, teachers, curriculum developers and workplace trainers.

The authors are well placed to undertake this exploration. John Keogh's 2013 PhD research "Looking at the Workplace through Mathematical Eyes" frames the book, which is further informed by his research for the Irish Government on work-based learning. Terry Maguire is an ideal co-explorer and co-author as the originator of the 'Maths Eyes' initiative which has achieved national and international prominence as a means of bringing mathematics to life and building confidence in an individual, everyday use of mathematics (www.haveyougotmathseyes.com/). 'Maths Eyes' is based on the idea that:

- Adults believe that the real world mathematics they can do in life and work is just 'common sense' the real world mathematics they can't do is mathematics.
- Having Maths Eyes changes the view of mathematics as being something everyone does in school. With maths eyes people see the mathematics they do, everyday mathematics becomes real and meaningful.

The spirit of 'Maths Eyes' pervades this book, in the belief that knowing what you know, as distinct from what you do, can facilitate recontextualisation for change (Keogh, Maguire, & O'Donoghue, 2014).

The authors are active in Adults Learning Mathematics (ALM) (http://www.alm-online.net/) the international research forum bringing together researchers and practitioners in adult mathematics and numeracy teaching and learning in order to promote the learning of mathematics by adults. As Founding Chair and an Honorary Trustee of ALM I am proud to announce that ALM celebrates its silver anniversary in 2018. Over the past 25 years it has provided a home base for explorers in this developing field, including the authors of this timely and important book. I commend this book to all those concerned with adults learning – and using – mathematics, in the workplace and beyond.

REFERENCE

Keogh, J., Maguire, T., & O'Donoghue, J. (2014). A workplace contextualisation of mathematics: Measuring workplace context complexity. Knowing what you know, as distinct from what you do, can facilitate recontextualisation for change. *Adults Learning Mathematics – An International Journal, 9*(1), 85–99.

ACKNOWLEDGEMENTS

This work would not have been possible without the cooperation, goodwill and encouragement of a lot of people, too numerous to name individually. Standing on the shoulders of giants is relatively easy. It is negotiating the climb that is arduous and challenging and where your support was most telling. Foremost among our support stands Ciarán O'Sullivan who was instrumental in adapting the tools that underpinned the research into practical, useful supports for teachers and learners. We are very grateful to the CEOs of Company A and Company B for permitting us generous access to their workplaces. A simple 'Thank you' seems unequal to the task of expressing our gratitude, but it is profoundly felt and succinct.

FIGURES AND TABLES

FIGURES

TABLES

CHAPTER 1

INTRODUCTION

What This Book Is About

People know more about mathematics than they think they do and use it more than they realise, especially in work. That they more readily account for what they 'do', dismissing what they 'know' as common sense or 'just part of the job', extends a self-perception of not 'being a maths person'. Such a mistaken self-perception can be transmitted within families, across generations and throughout communities as something that is inherited, inevitable and completely excusable, to an extent that literacy is not. At the same time, the pervasiveness of modern technology and the proliferation of information represented in mathematical ways is embraced by people with relative ease. This is the paradox lying at the heart of this work as it indicates that the reported disconnection with mathematics is rooted in self-perception and not due to some innate inability alone.

This book offers an explanation of this mathematics use/denial paradox, and a methodology for measuring instances of mathematics invisibility. It describes how people can recognise their mathematics competence, which they may dismiss as 'just part of the job', given the appropriate tools. It provides a lens through which any workplace (or social activity) context can be examined to identify and optimise work-based learning opportunities.

The extensive research which underpins this book was completed as doctoral research 'Looking at the Workplace through Mathematical Eyes' by the lead author (Keogh, 2013). The aims of the research were to identify and align the mathematics that people actually 'do' in work however unconsciously, with the formal provisions of the National Framework of Qualifications in Ireland in order to heighten mathematics visibility (NQAI, 2003). Readers who are interested in the full details of the research methodology employed and technical analyses completed are asked to refer to the thesis document for further information. While the research work was carried out in Ireland, the enabling strategies, analytical tools, and methods recognise and cater for the uniqueness of workplaces and their contexts, wherever they occur on the international landscape, and could be tailored to local conditions with relative ease.

This book discusses the findings of the research including those from a National Survey of People at Work which was validated by several case studies conducted to measure instances of Mathematics Invisibility. The resulting workplace Context Complexity Protocol tool that emerged can be used to enhance confidence in prior

 DOI:10.1163/9789004381766_001

acquired learning of mathematics and other topics, and enable its continuous re-contextualisation for employability and re-employability in the long term. A number of additional exploratory tools that were developed/adapted to investigate instances of mathematics in work that were otherwise invisible to the workers, and to map them on to the national standards framework will also be presented.

These tools can be self-administered, or used with the guidance of practitioners, to realise the following benefits:

- Reluctant learners can identify and acknowledge what they know about mathematics, however this knowledge was acquired or talked about, and establish a starting point at which formal learning can begin and be encouraged
- Experienced professionals, for whom mathematics is a qualifying component, could be helped to explore and clarify their practical applications of classroom-mathematics which have long slipped into invisibility. The exploratory tools can be used to peel away the obscuring layers of the ‘real world’, and to create a library of schemata for teachers to contextualise the meaning of, and justification for, elements of the curriculum
- Teachers can be informed of the contexts in which students’ prior learning occurs and can be exploited as a familiar platform to make meaning of mathematics concepts, while capitalising on the ‘real world’ schemata to situate explanations and practical applications for an enhanced learning opportunity.

The emergence of a workplace contextualisation of mathematics featuring a range of attributes that are common in varying degrees in every workplace, but entirely absent from explicit outcomes that characterise formal learning provision, is described. Furthermore, the paradox that while employers value experience gained tacitly on-the-job, they privilege formal accredited learning in recruitment, is highlighted, yet, having selected suitably experienced candidates, they proceed to re-train them in ‘how we do things around here’.

The authors argue that the National Framework of Qualifications in Ireland, while effective in its role of setting and maintaining standards for formal learning, could be extended to account for the range of workplace competences that are highly valued by employers yet acquired by tacit, informal and non-formal means. In this way, formal learning curricula could be enhanced to include workplace competences and improve work-readiness, while serving to encourage those with workplace competences alone to reinforce their knowledge with formal accreditation. Finally, a methodology that helps individuals to recognise the extent of what they ‘know’, that underpins what they ‘do’, serves to enable the re-contextualising of such knowledge to meet new purposes for employability and re-employability in an increasingly dynamic labour market.

This work is important because it seeks to make the use of mathematics in work more visible in formal terms, and to enable a reviewed self-perception as being mathematics-capable and all that that implies for long term employability and wellbeing. It is timely because it coincides with widespread concern, nationally and internationally, regarding mathematics ability, engagement with STEM education, and failing levels of performance in international comparative studies. It is ground-breaking because it attends to the topic in a novel combination of methodologies that were especially designed and adapted for purpose.

GUIDE TO STRUCTURE OF THE BOOK

Chapter 2 explores the nature of work and explains the tangled relationship between people and the mathematics-containing processes and procedures that enable their desired work outcomes. Such mathematics-saturation of the workplace context can be disguised in many ways perhaps confirming their self-perception of 'not being a maths person'. This chapter draws attention to mathematics at work and how it can be invisible to the worker. It explores the surrounding issues and highlights the embedded paradoxes and misconceptions, and their possible implications for workers.

Chapter 3 describes and explains the ways in which the otherwise invisible mathematics in work was identified, measured and calibrated for complicatedness in relation to the provisions of the National Framework of Qualifications.

Chapter 4 broadens the scope of this work by listening to what people in work had to say about their recognition and awareness of their numerate behaviour and their perception of its relevance to them, personally. It discusses the National Survey of Adults in Work and presents a summary of its key findings, which are surprising in what they report and the breadth of their implications.

Chapter 5 showcases four instances of people in the performance of their work observed through a mathematics-sensitive lens. It compares and contrasts their observed with their self-reported numerate behaviour as a measure of their mathematics invisibility. While each case study is different, they display thematic similarities which, while not generalizable, alerted the authors to their importance and their corroboration and authenticity.

Chapter 6 draws together the broad findings of the National Survey and the depth of the case studies and highlights the insights that enabled the development of a set of tools to support practitioners, teachers, and individuals as they face new challenges that new horizons will present. In the light of the evolving Fourth Industrial Revolution and the anticipated demise of the 'job for life', this work provides a platform to manage change in a way that captures existing competence and facilitates its re-purposing towards new, and as yet unknown, outcomes.

CHAPTER 2

ADULTS USING MATHEMATICS IN THE WORKPLACE

It's Not Mathematics, It's Just Common Sense; Part of the Job

The idea that adults use mathematics in the workplace is not new. Mathematics activity is typically associated with contexts that are number-explicit and none other. This chapter shines a light on the nature of work to reveal the contrast of what is readily observed and assumed, with features that operate to conceal and obscure. The following sections consider some of the factors that complicate the relationship between adults at work and how, through their numerate behaviour, they display their affinity with mathematics-saturated activities that extend beyond quantity and number to pattern and relationship, space and shape, data handling and chance.

The term 'Mathematics' is commonly used in a broad range of contexts and discourses, often without qualification as to type, purpose or level of difficulty. It can connote particular collections of related concepts, tools, symbols and techniques ranging from the most basic, graduating in degrees of complicatedness towards the abstract and theoretical, concerned with the extreme frontiers of research and development. Mathematics may be understood as an intellectual discipline, important for its own sake, for the knowledge economy, for science, technology and engineering, for the workplace and the individual citizen (Smith, 2004). In this context, mathematical skill is much sought after for its power to place logical and critical reasoning at the service of problem solving.

'Mathematics' can evoke recollections of school curricula and may be accompanied by negative feelings and anxiety, rooted deeply in the early years education and persisting beyond the classroom (Klinger, 2008). However, real life and work situations are not neatly packaged and appropriately labelled as to the relevant mathematical discipline which will provide solutions. Instead, workers approach practical tasks with a broad range of skills, attitudes and knowledge including:

- Situational (vocational and other) knowledge and skills, (in particular, mathematical);
- Meta-cognitive skills and strategies (e.g., critical thinking, planning, problem solving and evaluating);
- Personal skills (Fitzsimons, 2002).

 | DOI:10.1163/9789004381766_002

The mathematics of interest here is equated to the Mathematical Knowledge, Skills and Competence (MKSC) that underpin and are subsumed by the performance of work, as a means to an end, external to its theoretical provenance or means of acquisition (Fitzsimons, 2013). They are acquired in formal, non-formal and informal ways in response to a work-based problem, and cannot not be encapsulated by a single convenient label that is easily interpreted by reference to a curriculum, and maybe referred to as 'numeracy'.

For some, numeracy encapsulates a range of knowledge and skills needed to be able to manage and respond to the mathematical demands of diverse situations (PIAAC, 2009). Common mathematical themes encompassed by numeracy stem from van Groensteijn's proposal in 2002 (van Groensteijn, 2002), later refined by Maguire as including, Quantity and Number, Space and Shape, Pattern and Relationship, Data handling and Chance, and Problem Solving (Maguire, 2004), and adapted for PISA 2003 (OECD, 2004) and subsequent International Surveys. This segregation of mathematical domains has been carried through to the official framework in Ireland with which mathematics learning by adults is accredited at the lowest levels. 'Mathematics' and 'Numeracy' are sometimes used in the literature to describe the same field of study (Fitzsimons, Jungwirth, Maasz, & Schloeglemann, 1997). The latter has been differentiated from the classroom discipline and has come to refer to the range of mathematical knowledge and skills that adults use in managing situations in their everyday lives (NALA, 2004). Numeracy involves cultural, social and emotional facets (Maguire & O'Donoghue, 2003), and enables effective participation, albeit bounded by the context in which it is acquired (Evans, 2000a). In a dynamic world, mathematical habits of mind i.e., discussing, sharing, challenging and reflecting, are an ever-present feature of the modern workplace and support the individual in responding to change (Hoyles, 2008).

The notion of workplace defies definition other than in the most general terms e.g., being engaged in an activity, not necessarily dependent on time or place, for a consideration which is material or otherwise. In the modern workplace, skills deployed in the performance of work lie at the kernel of a number of encapsulating layers, each exerting its own influence to hide, wholly or partially, the underpinning MKSC. The workplace comprises chains of activity that are triggered in different combinations. Activities encapsulate work practices, which themselves are a combination of tasks, normally considered to be a 'job'. Underpinning a 'job' is a collection of knowledge, skills and competence, the mathematical content of which lie under the cover of a range of factors, whether internal, external, personal or contextual (Figure 2.1).

Not only are MKSC obscured from the view of Management and Operator alike, albeit in varying degrees and combinations, the performance of the tasks upon which they rely, are buried deeply within organisations. References to a job usually denote a person, engaged in a work practice alone or in concert with others to maintain workflow from initiation to completion.

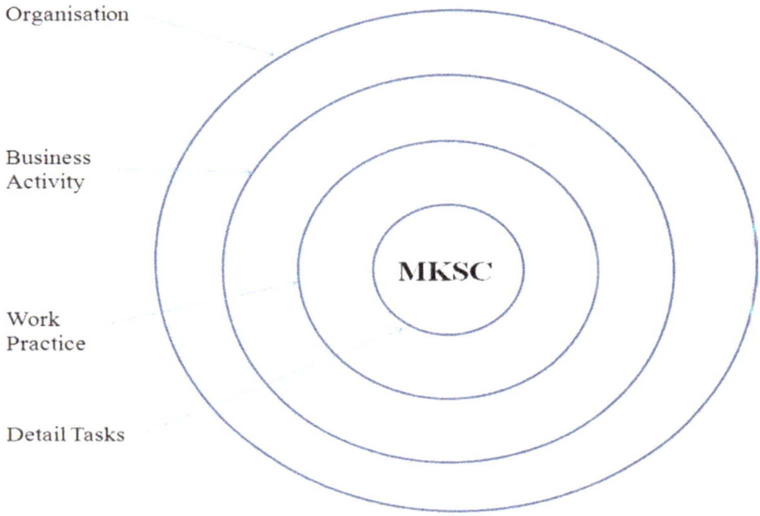

Figure 2.1. Mathematics knowledge skills and competence (MSC) surrounded and concealed by multiple layers

In the mid-1970s, Benoit Mandelbrot coined the term 'fractal-like' to encapsulate the idea that the world is full of self-similar structures, which, while rugged at one level of magnification, can display ever finer yet similar detail, in another. To the uninitiated, an organisation appears to be an entity characterised by a single culture, purpose and population of homogeneous employees, engaged in activity to produce goods or provide a service. Closer examination reveals that the organisation is divided into specialities, each with particular artefacts to support their activities, governed by formal and informal rules, populated by a community with their own priorities, self-images and expectations, such that may bring them into conflict with other centres of speciality. A more magnified view uncovers that individuals, or groups of individuals within a speciality have developed their own artefacts, culture, self-image, and norms that do not necessarily share those of the speciality to which they 'belong'. Contemplating the individual brings unique factors to the surface which are similar, but not identical to that of the group, within the speciality and within the organisation. Work shadowing of individuals highlighted that the fractal perspective persisted through to the performance of work. Rather than complete a series of tasks in an uninterrupted sequence from initiation to completion, the worker continuously switched between a variety of activities, each differing in aspects of artefact, rules, community, division of labour, status, and meaning. This insight enables an explanation of the workplace mathematics use/denial paradox from a Cultural Historical Activity Theory perspective (Figure 2.2).

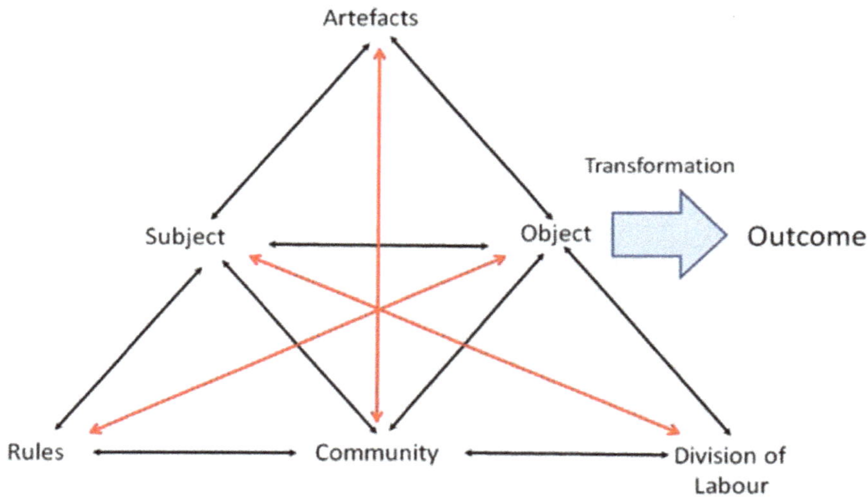

Figure 2.2. Co-dependent and inter-dependent affective factors or nodes of a social activity

CULTURAL HISTORICAL ACTIVITY THEORY

Cultural Historical Activity Theory (CHAT) is widely used to describe the interdependent constituents of human activity. CHAT, first graphically illustrated in the 1980s (Engeström, 1987, 2001; Engeström & Cole, 1997; Engeström, Miettinin, & Punamäki, 1999; Leont'ev, 1978; Roth & Lee, 2007), posits that human activity comprises subject(s), gathered in a community, who form and are formed by rules, among whom labour is divided, and who are joined in common cause in pursuit of an object or outcome, even if they are differently motivated (Figure 2.2).

Workers (subject) who are joined in common cause in a work group for example, are drawn from a community among whom labour is divided. They use artefacts, i.e., knowledge, skills, competence, language, tools etc., to achieve an objective which may not be completely shared but results in an outcome. They engage in activities that are constrained by formal, explicit rules and shaped by informal rules which are often unspoken. The anticipated flow of work is often disrupted by events, e.g. breakdown, that bring workers into encounter with other work activities each with its own rules and objectives and so on. This adds tiers of complexity to the workplace that can go unnoticed with extended exposure, but which must be integrated into work practice nevertheless. That workers behave in numerate ways, regardless of the context, tends to associate mathematics with multiple contexts subliminally but none explicitly, which subsumes their recognition of mathematics in favour of their work competence.

COMPETENCE

Competence, especially in relation to workplace, is a complex concept (Fitzsimons, 2002). It is sometimes used interchangeably with *competency* i.e., as an indicator of skills performance, and is task oriented. Competent performance or effective action implies the mobilisation of knowledge, cognitive and practical skills, as well as social and behaviour components such as attitudes, emotions, values and motivations, according to the OECD report on The *De*finition and *Se*lection of Key *Co*mpetencies (DeSeCo) 2005. Since workplace functions are constantly changing in response to market forces, so too will the relevant competencies. Furthermore, DeSeCo reports that each competency is not used in isolation, and each context may require a "constellation of competencies" that changes in response to changes in technology and in social and economic structures (Rychen & Salganik, 2005).

This is a more comprehensive view insofar as it reaches beyond the practical application of knowledge and skill, indicating their creative and effective deployment in human situations defined in terms of Context, Role, Learning to Learn and Insight.

Context: This dimension of competence is characterised by the complexity of human situations and correlates with diversity. The contexts in which knowledge and skills may be applied, are constructed by humans and range from the simple and very structured at one end of the spectrum to the highly complex and unpredictable at the other. There is research suggesting that many 'jobs' consist of quite basic mathematics being deployed in complex situations, or that much complex reasoning is required before the application of basic mathematical operations, exemplifying lower levels of knowledge being deployed in ways that indicate higher levels of competence (Wadsworth, 1997; Wake, 2005).

Role: The variety of *roles* occupied by a worker ranges from the simple, highly defined and supervised requiring lower levels of learning, to the diverse, volatile and unpredictable requiring the application of social and leadership skills associated with higher levels of learning. Here too, more advanced, and perhaps multiple roles taken on by a person with elementary levels of knowledge, are evidenced by the individual's ability to adopt the appropriate role on joining a group whether simple or diverse.

Learning to Learn: The ability to recognise one's limitations and taking the appropriate measures to transcend them, within the structures provided is captured within this dimension of competence. To observe and participate in new experiences, extract and retain new meaning from them, while drawing on other stands of knowledge and skill, underscores the relationship between learners and their learning processes. At the lower end of the spectrum, potential learners may not attend to their learning needs because they are unaware of their existence (Maguire & O'Donoghue, 2003).

Insight: The levels of competence linked to *Insight* imply a scale of understanding and consciousness developed by reflecting on experience and feedback from the general environment. The ability to integrate other learning outcomes with one's set of attitudes, motivations and beliefs, enhances self-understanding and promotes increasing levels of autonomy in interaction with society.

Where the dimensions of context and role in the classroom are, by and large, fixed by history and tradition, the workplace tends to be a more dynamic environment, especially nowadays. Where the syllabus and assessment criteria are known, the measurement of learning outcomes can be relatively straight forward. In contrast, the workplace varies in response to, and in interaction with, a wide variety of influences ranging from the personal to the organisational such that knowledge and skills assessed at one level, may need to be augmented by different levels of competence influenced by dimensions formally defined as Context, Role, Learning to Learn and Insight.

Mathematics in work lives in a complex space that speaks of a sophisticated environment in which even the most basic of mathematics concepts are realised. Workplace contexts are unique by virtue of being populated by people who are unique and who interact in unique ways. They are made complex by varying degrees of accountability, clarity, familiarity, stress and volatility. The prominence of what needs to be 'done' focuses on actions and relegates the visibility of the enabling mathematics for a number of reasons.

INVISIBLE MATHEMATICS IN THE WORKPLACE

In the workplace, the visibility of mathematics or otherwise is a combination of relationships between the person, their self-perception and the nature of their work; but it is not as simple as that. Although current discourse is replete with concern regarding the importance of STEM subjects to the wellbeing of future generations, it is often counterbalanced by statements made by public figures claiming to have been 'rubbish at maths', which gives the impression that mathematics is only accessible to certain types of people. Such self-perception, however misguided, can normalise the idea that one can either 'do' mathematics or not, and explains their numerate behaviour as common sense or anything but mathematics. The following paragraphs outline the factors that shape an attitude to mathematics in general and in work and how mathematics might become invisible as a result.

Insufficient mathematical knowledge, skills and competence can have a negative impact on a person's wellbeing and employment prospects in the long term (Bynner & Parsons, 1997, 2005). Furthermore, anxiety about mathematics and poor outcomes in mathematics learning, can be explained by individuals as an unavoidable, pervasive and permanent human characteristic of 'not being a maths person'. Such attitudes and beliefs about mathematics limits educational choice, constrains development opportunities, and can be transmitted within and across communities.

The self-perception of 'not being a maths-person' is a function of the public image of mathematics as being remote and alien to personal and professional lives, (Ernest, 1995), while socially important, mathematics may be subjectively irrelevant and invisible (Niss, 1994; Wedege & Evans, 2006), tending to explain the mathematics content of their work as 'just part of the job' or common sense (Harris, 1991) or anything but mathematics (Coben, 2000). A negative self-image in relation to mathematical ability, promotes a common sense explanation for any mathematical activity displayed (Marr & Hagston, 2007). Common sense is an innate human characteristic which is motivated by the desire to behave intelligently. It is developed and maintained by a cyclical guide to action viz., experience, analyse, decide, execute and is unique to each individual depending on their cultural context (Colleran & O'Donoghue, 2007), (Colleran, O'Donoghue, & Murphy, 2003). It can accumulation of situation schemata that gather with experience for later recall in encounter with similar events to inform their actions (Colleran et al., 2003). It is resilient as a matter of belief, with no explanation necessary and is a *pot pouri* of disparate notions applied *ad hoc*, and therefore is in sharp contrast with the rigorous application of mathematical methods (Geertz, 1993). That it is necessarily perceived as good sense, and something to be transcended rather than rejected' (Gramsci, 1971), inspired Coben to recognise the power of lived experience as an educative process that could be harnessed by adults learning mathematics (Coben, 2002). In contrast, more visible mathematical habits of mind, i.e., discussing, sharing, challenging and reflecting support the individual in adapting to change in work practice and facilitate employability in the long term (Hoyles, 2008).

In addition, MKSC learnt in a specific context, for a specific reason e.g., so-called ethno-mathematics (D'Ambrosio, 1985), can be bound by that context and unavailable otherwise. Similarly, MKSC can be embedded in artefacts designed by others, tacitly learnt and intertwined with processes and procedures (Hoyles, Wolf, Molyneux-Hodgson, & Kent, 2002), which construct and reinforce a discontinuity between a school understanding of mathematics and that which underpins workplace activity, leading to fragmented recall that is difficult to transfer, if not impossible, without some inter-discourse translation (Evans, 1999, 2000a).

On-the-job training compounds mathematics invisibility by being focused on what to 'do' rather than 'need to know' and reflects employers' value judgments rather than demonstrable fact (Hoyles et al., 2002; Marsick, 1997), relying on the workers to develop multiple problem solving abilities rather than a finite range of algorithms suggested by formal school mathematics (Hudson, 2008).

The 'Recency' effect, which echoes the 'use or loose it' aphorism, refers to the inverse relationship between skills and age ascribed to all countries (Rychen & Salganik, 2003), whereas the lack of opportunity to practise skills accelerates their gradual obsolescence over time contributing to the mathematics invisibility effect (Coben, 2005). The appearance of gender bias of facility with mathematics

(Chipman & Wilson, 1985) is thought to deter girls, reinforced by the concept of '*habitus*' impinges also, not least because of its durability (Bourdieu, 1977; Wedege, 1999) and its power to excuse a lack of MKSC.

Denial of one's mathematics ability, as a consequence of a bad experience can persist over many years and permit negative attitudes to project across generations (Benn & Burton, 1993). Access to mathematics discourse, crucial for the acquisition of MKSC, can be prevented or limited for want of appropriate language and terminology (Wake & Williams, 2007).

Status within a group or Community of Practice (CoP), moderates the visibility of the embedded mathematics. For example a newcomer may display a particular facility with recently introduced artefacts, whereas a longer-serving worker's knowledge may be rooted in experience and informed judgement (PIAAC, 2009). According to Wenger, members of a CoP have resources, history, discourse, and problem solving methodologies in common which encapsulate their practice and bind the community together (Wenger, 2006) and propagates an acceptable perception of mathematics whether overt, covert or otherwise. Similarly CoPs feature 'tricks of the trade', or 'the way we do things around here', that envelop discourse, heuristics, algorithms, implements or badges of achievement and history with a similar effect on the visibility of the underpinning mathematics (Engeström & Cole, 1997).

The development of highly sophisticated artefacts is gradually, imperceptibly and progressively hiding mathematics from view such that mathematical skills are a last resort in non-routine situations e.g., when the expected flow is disrupted (Strasser, 2003).

Mathematical structures encountered in the workplace although simple require profound understanding and mathematical thinking to develop a complete understanding of its meaning (Wake, 2005). ICT devices embed mathematics in work practice in such a way that the user need only be aware of the extent of their interaction with such devices, leaving the technical specifics to the specialist. As a result ICT is often referred to as 'black boxes' implying the 'plugging–in' of variables to generate a specific outcome, with little or no understanding of the mathematical content or activity (Williams & Wake, 2007). The task to supply certain quantities, from specified sources, to feed an algorithm for interpretation by another 'job', can require sophisticated mathematical reasoning and processes of which the worker is unaware. It has been argued that, rather than compounding mathematics invisibility, understanding such algorithms could facilitate meaning making for the worker and enhance fulfilment and reflect workplace culture (Hoyles, Noss, Kent, & Bakker, 2007).

There is a catalogue of factors that exacerbate mathematics invisibility e.g., Language/Jargon (Wake & Williams, 2007); Habitus (Bourdieu, 1977; Wedege, 1999), Training (Brown et al., 1989); Common sense (Coben, 2000a; Colleran & O'Donoghue, 2007); Group Status (PIAAC Numeracy Expert Group, 2009) (Lave & Wenger, 1991) (Lave & Wenger, 1991); Artefact (Marr & Hagston, 2007;

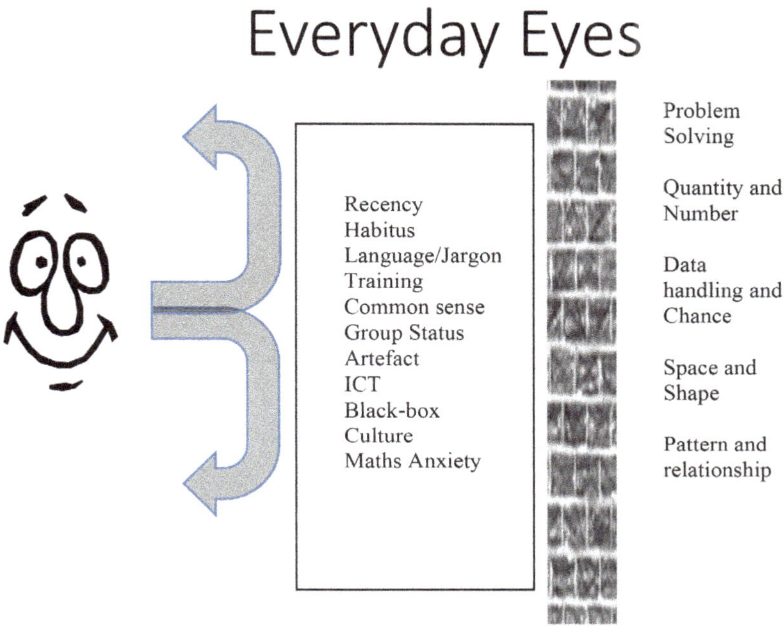

Figure 2.3. Mathematical knowledge, skills and competence obscured by workplace factors

Strasser, 2003); Black-box (Hoyles et al., 2002; Wake & Williams, 2007); Culture (Colleran et al., 2003) and Anxiety (Evans, 2000b; Klinger, 2008; McLeod, 1992). These operate in different combinations, varying from one individual to the next, to obscure or deny such skills, however unwittingly. Just as human vision perceives only a narrow spectrum of light, so too will many overlook the mathematical concepts that pervade everyday work and life. Seeing things through what might be called 'Everyday Eyes' gives prominence to what workers 'do', rather then what they 'know', and compounds the self-perception of not being a 'maths person' and constrain the potential for adapting to change and maintaining continuing employability (Figure 2.3).

In contrast, viewing the world through what is called 'Mathematical Eyes' perceives problem solving, space and shape, pattern and relationship, data handling and chance, in addition to quantity and number (Figure 2.4).

The concept of 'Mathematical Eyes', inspired by Amy Benedicty (Benedicty & Donahoe, 1997) and developed by Maguire (Maguire, 2003, 2012), encapsulates the sense that humankind experiences the 'real world' everyday, which incorporates many mathematical themes. Rather than being a separate, isolated, absolutist (Ernest, 1995), classroom-bound phenomenon, the many facets of mathematics are

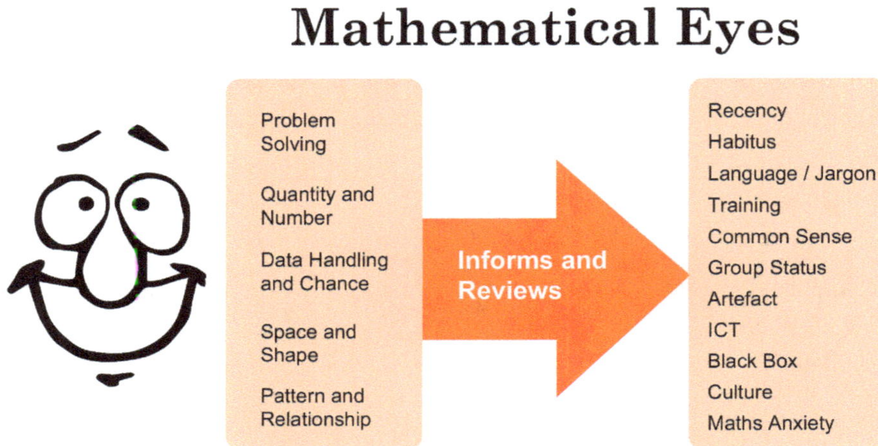

Figure 2.4. Visible mathematical knowledge skills and competence

inextricably bound up with the real world, made visible once having developed the eyes with which to see them.

Visible MKSC elements, enhance an understanding of workplace factors and establish a platform on which to adapt, change, innovate and create. Visible MKSC can be exploited to develop situation specific mathematical procedures and know-how, as reported by the PIAAC Expert Group (PIAAC Numeracy Expert Group, 2009).

Opening Mathematical Eyes is problematic, especially in the workplace where tasks become interwoven into an amorphous whole, i.e. the 'job'. That mathematics could become so inextricably bound up in one's work practices as to be invisible presents different yet interrelated problems for different stakeholders. Employers, for whom a profile of the skills requirement of a particular job overlooks the MKSC elements, can struggle to recruit and train personnel. Workers that are mismatched to their jobs are less likely to realise their potential, be positively responsive to change, and to obtain job satisfaction which negatively impacts on retention and absence (McClelland, 1988). Moreover, the tendency to dismiss such expertise as mere 'common sense' impedes the Recognition of Prior Learning (RPL) (Coben, 2000b). Providers of workplace learning opportunities are constrained in the design of programmes for the want of a starting point, especially for topics that are not seen to be pertinent to the job (Klinger, 2009). The extent of mathematics invisibility experienced by individuals in the workplace, can be a function of the way they engage in their work activity and how they talk about it (Keogh, Maguire, & O'Donoghue, 2012; Keogh, Maguire, & O'Donoghue, 2013). The latter can be a function of the discourse that is particular to and pervades the workplace.

The discourse of workplace activity, features words, expressions, references, abbreviations and acronyms, drawn from a common language, but bearing meanings particular to the participating community. The presence of mathematics concepts and ideas may not be acknowledged for activity theoretical reasons but referred to in ways that are coherent to the subject(s) of the activity system alone (Keogh, Maguire, & O'Donoghue, 2016).

Bernstein provides a useful framework to support the analysis of discourse and how knowledge is developed, accumulated and circulated, under conditions determined by the particular context in which it occurs (Bernstein, 1999). Developing an understanding of the nature and substance of local discourse contributes an explanation for the denial/use of mathematics in workplace settings. According to Bernstein, discourses are differentiated by the structure of the knowledge it has been developed to support, and the ways in which that information is shared, excluding all but the initiated. This view accommodates both formally and informally acquired knowledge, its structures and their attendant specialised languages.

Workplaces are replete with words and phrases that take on particular and local meanings. For example, a case study participant describes a situation where what transpired to be the optimising of the available space in a warehouse in three dimensions was called a "Tetris job" (Keogh, 2013). This referred to a 1980s computer game that invites the player to fit the maximum number of irregular 2D shapes into a confined space and under the pressure of time. In another activity, 15 digit codes were used, wholly and partially, to generalise, specify, store, retrieve, forecast and communicate in a way that is inaccessible to the observer without explanation or translation.

It is likely that mathematics activity, at once invisible, unconsciously performed, denied through self-perception, referred to in terms understood by the activity system's subjects/community alone, compounds the difficulty for the "outsider" and detected only by methods of enquiry capable of transcending these obscuring barriers (Williams & Wake, 2007). The workplace activity is designed and performed with the object of getting a piece of work done repeatedly, in time, and to fairly explicit specifications, whereas the researcher seeks to uncover MKSC that is concealed in some way. In the case studies, the worker's introspection was forced in response to enquiry or process breakdown and compromised as a result. Communication between work colleagues features truncated references which carry loaded meaning for them but remain beyond the reach of the researcher. Explanations, when given, are at best translations for the researcher, shrouded in fear and suspicion of the "outsider", and depend on the skill and willingness of worker to articulate and the researcher to comprehend. Similarly, being invited to explain how a decision is arrived at, based on the balance of probability, elicits a response expressed in terms of experience and intuition, e.g. 'you just get to know'.

In sum, there are many background and foreground factors that bear on individual workers in unique combinations and degree. Workers encounter their workplace, which itself is unique due to the confluence of organisational cultures, history, ethos, processes and procedures, each of which have several informal sub-versions, and the presence of other unique individuals (Skovsmose, 1996). While the purpose at the core of the workplace is familiar and previously experienced elsewhere, the individual's accumulated MKSC is rendered invisible by the assimilation of actions, processes and procedures designed to meet the particular requirements of the new context without reference to underpinning knowledge and skills.

Mathematics activity in the workplace is not conveniently packaged and signposted in line with a mathematics framework, nor is it concerned with the pursuit of "deeper mathematical understandings" (Fitzsimons, 2005). It is hidden by a number of layers of activity systems, their surrounding discourses, learning methods and, crucially, the expectations, attitudes and beliefs of the worker (Keogh, Maguire, & O'Donoghue, 2010b). The extent of mathematics invisibility is a subjective experience, unique to each individual and established by reference to a system of measurement and in comparison to other workers. In addition, rather than identify and calibrate examples of the application of overt mathematical operations as might be associated with engineering or actuarial activity, many of what may be considered low skilled jobs require mathematical ability expressed as numerate behaviour.

Numerate Behaviour is defined by the OECD Expert Numeracy Group as responding to mathematics containing information, represented in a number of ways (PIAAC Numeracy Expert Group, 2009). These include 'Identify', 'Evaluate', 'Act upon' 'Communicate about' and 'Problem Solve'. The type of information evoking the numerate response contains mathematics characterised as belonging to one of four domains, viz., Quantity and Number, Space and Shape, Pattern and Relationship and Data handling and Chance. These domains mirror the curriculum framework which presents mathematics at the lowest levels of accreditation for adults. The authors readily acknowledged that arbitrarily separating mathematics concepts and ideas into sharply defined domains, while convenient for curriculum design in the short term, may ultimately obscure the connectedness of the 'big ideas' of mathematics. At the same time, seeking to disentangle the role of mathematics in the service of work, benefited from an approach that broke down processes and procedures into their components, by identifying specific numerate behaviours in conjunction with mathematics domains that are officially recognised nationally and internationally.

Viewing the world of work through the innovative lens of 'Mathematical Eyes', the authors used a National survey to uncover the many layers that hide the mathematical underpinnings of the processes and procedures that support and comprise specific jobs and to measure the extent to which they may, or may not, be recognised formally, and by the workers themselves. To be clear, this approach is rooted in the identification and calibration, in formal terms, of the mathematics that workers were observed to have used in the performance of their work, for the benefit of their development, long-term employability and wellbeing. It did not intend to specify and remedy gaps in their mathematics knowledge, skills and competence (MKSC) from a curriculum perspective, nor did it intend to mathematise work to reinforce any particular point of view.

This work was informed by a survey for breadth, accompanied by case studies for depth of understanding the perceived role of mathematics in work. The purpose of the National Survey of People at Work in Ireland, Appendix 1, was to capture the extent to which individuals recognised the detail of their custom and practice through the lens of their numerate behaviour. The findings of the survey were supported by comparing individual workers self-reported numerate behaviour with that observed by job-shadowing case studies, thus enabling their mathematics invisibility to be measured.

CHAPTER 3

IDENTIFYING AND MEASURING MATHEMATICS INVISIBILITY IN THE WORKPLACE

Stripping away the Layers That Hide What a Person Knows about Mathematics

The different characterisation of mathematics in the workplace, offers a platform upon which to temper mathematics learning and curricula-design in preparation for life and work. More than ever, problem solving, spatial awareness, estimation, interpretation and communication skills, are highly valued in the modern worker, as being essential to support change, reaction and response especially given the pervasiveness of ICT and 'black boxes', whether technological or algorithmic (FÁS, 2009; O'Donoghue, 2000).

From the perspective of the individual, the Mathematics, knowledge, skills and competence (MKSC) that underpin work, whatever its category, type or domain, may be dismissed as 'just part of the job' (Coben & Thumpston, 1995). Skills deployed from a 'common sense' perspective tends to conceal mathematical ability rather than expose it for development (Coben, 2002). The paradox that workers may deny their use of mathematics while relying on them in work, unjustifiably extends the self-perception of 'not being a maths person'.

To dismiss mathematical skills, rendering them invisible, poses significant challenges for education and training programmes for want of a starting point, i.e. the so-called 'bootstrap problem' (Klinger, 2008). This is problematic for the worker, since, in the view of the National Development Plan (NDP) in Ireland, a lack of appropriate MKSC can have a critical impact on a person's continuing employability. Longitudinal studies, conducted in the United Kingdom, tend to confirm the correlation between low levels of numeracy and poor long term employment prospects, socio-economic deprivation and wellbeing (Bynner & Parsons, 1997, 2005; Williams, Clemens, Oleinikova, & Tarvin, 2003). This is further underlined by the report of the Expert Skills Group in Ireland, that persons with low levels of education are at the greatest risk of unemployment, only 5% of all vacancies being for lower skilled jobs (FÁS, 2009). Accordingly, the penalties attached to the absence or insufficiency of mathematical knowledge, skills and competence are quite severe.

While the effect of the absence of MKSC and its being invisible to the individual are similar, an understanding of mathematics invisibility in the workplace offers the possibility of making it visible and available to the individual for development.

 | DOI:10.1163/9789004381766_003

Workplace recruitment and training procedures can benefit from recognising the depth of anxiety and reluctance to engage that some employees experience, in anticipation of induction and change. Such a range of recognition requires the adoption of a different lens through which the workplace may be observed.

There is the widely reported phenomenon of "Mathematics Invisibility" thought to be especially prevalent in the workplace. While some workers perceived themselves as 'not being a maths person', routinely denying their use of mathematics or dismissing it as something else, they nevertheless acknowledged their Mathematics Knowledge Skills and Competence (MKSC) in specific settings in their responses to the survey items (Chapter 4). There was a range of factors that influenced the perceived role of mathematics in the workplace which, not being accounted for, tended to compound 'Mathematics Invisibility'. It was not just a question of how complicated the mathematics knowledge and skills underpinning work were, but also the complexity of the situations in which they operated.

Traditional methods for identifying and measuring facility with Mathematical Knowledge Skills and Competence (MKSC) are based on the schools assessment approach in which mathematics problems are contrived to test recall of a curriculum. To do so in the workplace would catastrophically compromise the attempt to measure mathematics in work, instead transporting the individual to, perhaps negative, memories of the school mathematics classroom (Maaβ, 2005). International surveys, while recognising this likelihood, are based on attempts to design curriculum-independent, 'real world' tests, which, despite the best of intentions, may not reflect the actual MKSC that underpin the performance of an individual's work.

Even though people may deny or dismiss their use of mathematics, to solve problems or manage mathematics containing situations they still behave in numerate ways that are underpinned by MKSC (PIAAC, 2009). This concept underlies the methodologies and instruments developed in this work to identify mathematics activity in the workplace, to describe its contextualisation, and to contribute an explanation for the impact Mathematics Invisibility has on the workplace.

A National Survey of People in Work (Appendix 1) completed in Ireland, explored numerate behaviour in the workplace and the beliefs and attitudes of those at work to mathematics. In particular the survey focused on the mathematics that may underpin what might be called low-skilled or relatively low-paid jobs that do not feature obvious displays of mathematical activity.

To make the MKSC visible the authors devised a range of tools that were used to profile the structure of the organisation methodically to strip away the occluding layers in the sequence in which they were encountered, to separate combinations of workplace activities and to identify their constituent jobs. Once identified the MKSC underlying the tasks, were classified as to mathematics domain, level of difficulty and situational factors, and benchmarked with the NFQ in order to communicate about them. The tools developed include:

- Business Activity Modelling (Appendix 2)
- Work Practice Modelling (Appendix 3)
- Detailed Task Analysis tool (Appendix 4).

In practice, the application of these exploratory tools hinted at their potential to capture the circumstances in which the knowledge, skills and competence of a broad spectrum of topic disciplines, beyond mathematics, were expressed, especially in a fractal-like workplace (Chapter 4).

SURVEY METHODOLOGY

The National Survey of People at Work was designed by the authors to probe attitudes and beliefs around the awareness of, and the frequency of use of, mathematics in work. Following an extensive review of the pertinent research literature, the authors conducted a Pilot Study comprising three site visits over a period of four weeks. This provided an opportunity to field test prototype instruments and field research technique.

The survey was launched and remained open for three months. The sample achieved was large (N = 449 valid responses) and ranged across the full spectrum of mathematics education level, gender, and job skills level.

The typical interpretation of ordinal scale data was complemented by the application of the Rasch Measurement Theory to authenticate the survey items, to underpin their validity and triangulate the findings. Full details of the survey methodology and in-depth analysis is detailed in the original thesis (Keogh, 2013).

The National Survey of People at Work was prompted by the absence of relevant data from the field of study and is the first of its kind in Ireland. To date the workplace data that is available focuses on employment levels, gender representation, permanency, rates of pay and contractual matters by reference to industry sectors. They are often informed from the employers' perspective and as a statutory requirement.

The survey is built around the construct of Numerate Behaviour in encounter with specific mathematics domains, in order to facilitate a broad range of enquiry into differently nuanced workplace situations. In addition to demographics, survey participants were invited to report their Numerate Behaviour. There are many background and foreground factors that may bear on individual workers in unique combinations and degree (Skovsmose, 1996). Workers encounter their workplace, which itself is unique due to the confluence of organisational cultures, history, ethos, processes and procedures, each of which may have several informal sub-versions, and the presence of other unique individuals. While the purpose at the core of the workplace may be familiar and previously experienced elsewhere, the individual's accumulated MKSC are further rendered invisible by the assimilation of actions, processes and procedures designed to meet the particular requirements of the new context without reference to underpinning knowledge and skills.

Against this background, the application of traditional methods of identifying and measuring capability with mathematics faced significant challenges.

MEASURING MATHEMATICS CAPABILITY

Many countries across the World recognise the value of being able to compare and contrast their different education systems, their organisation and curricula to inform policy making. The OECD has taken a leading role by conducting a series of surveys concerning the literacy, numeracy and scientific attainment of children and adults in its member and other countries. The PISA series focused on 15 year old children in school, while the International Adult Literacy Survey (OECD, 2000), Adult Literacy and Lifeskills Survey (OECD, 2005) and the Programme for the International Assessment of Adult Competencies (Central_Statistics_Office, 2013) addressed the adult population ranging in age between 16 and 65 years. The Trends in International Mathematics and Science Study series (TIMSS), focussed on the effectiveness of teaching and learning mathematics and science in children in the 4th and 8th grades in school. These longitudinal studies provided the basis for comparison between countries competing for global inward investment, to gauge the efficacy of pedagogical interventions, and to inform policy makers. They also employed similar methods for measuring mathematics ability, by posing problems that require some mathematical technique to produce a solution. In choosing the mathematical/theoretical framework to develop the survey both PISA and PIAAC were considered as possible models. PISA is underpinned by a conceptualisation of mathematical literacy while PIAAC is associated with a conceptualisation of numerate behaviour. Both constructs contain mathematics or quantitative information, in real world contexts that require mathematical knowledge, skills and competence in their response. Both describe different and capture different levels of MSKC (Gal & Tout, 2014). PISA is more saturated with direct references to mathematics where PIAAC in contrast focuses on tasks that adults respond to in their everyday life communicated in a way that emphasises the tasks and what is required to manage them effectively. Further, the PIAAC numeracy construct is free of school-related considerations that underpin the PISA framework (Gal & Tout, 2014).

MEASURING MATHEMATICS CAPABILITY IN THE WORKPLACE

Following the lead of international surveys to assess mathematics use in the workplace in any absolute way is unlikely to succeed. Firstly, the research site would be catastrophically compromised by metaphorically catapulting the participant out of the workplace and back to the mathematics classroom. It could stimulate memories about mathematics learning and examinations in school that could range from positive to negative and perhaps some measure of examination anxiety. It would test powers of recall of curriculum and perhaps comment about the progressive deterioration of mathematical knowledge and skills over time. Most critically, such tests would be unlikely to uncover the actual mathematical underpinnings of the individual's performance of their work.

Mathematics activity in the workplace is not conveniently packaged and signposted in line with a mathematics framework, nor is it concerned with the pursuit of "deeper mathematical understandings" (Fitzsimons, 2005). It is hidden by a number of layers of activity systems, their surrounding discourses, learning methods and, crucially, the expectations, attitudes and beliefs of the worker (Keogh, Maguire, & O'Donoghue, 2010b). The extent of mathematics invisibility is a subjective experience, unique to each individual and may be established by reference to a system of measurement and in comparison to other workers. In addition, rather than identify and calibrate examples of the application of overt mathematical operations as might be associated with engineering or actuarial activity, this research observed what may be considered low-skilled jobs, where mathematical ability is expressed as numerate behaviour.

NUMERATE BEHAVIOUR

Numerate Behaviour is defined by the OECD Expert Numeracy Group as responding to mathematics containing information, represented in a number of ways (PIAAC Numeracy Expert Group, 2009). This work adopted 'Identify', 'Evaluate', 'Act upon' and 'Communicate about', as the principal responses. Problem Solving, as an enabling process, was omitted from the language of this case, as it bears multiple, not necessarily mathematical connotations for the workplace and workers. Nevertheless, workers necessarily solve problems in whatever context they arise, guided by their experience, acuity and work practice. The type of information evoking the numerate response may contain mathematics characterised as belonging to one of four domains, viz., Quantity and Number, Space and Shape, Pattern and Relationship and Data handling and Chance, as determined by the official mathematics framework in Ireland. Functions, insofar as they may arise at the level of this research, are thought to be adequately catered for by the Pattern and Relationships domain. The authors readily acknowledge that arbitrarily separating mathematics concepts and ideas into sharply defined domains, while convenient for curriculum design in the short term, may ultimately obscure the connectedness of the 'big ideas' of mathematics. At the same time, seeking to disentangle the role of mathematics in the service of work, may benefit from an approach that breaks down processes and procedures into their components, by identifying specific numerate behaviours in conjunction with mathematics domains that are officially recognised nationally and internationally. The findings of the survey are also supported by a number of in-depth, job shadowing, case studies carried out in the workplace. These are discussed in more detail in Chapter 5.

The intention to explore the possible impact of mathematics invisibility in the workplace suggests that such invisibility is a binary condition; either a person perceives, or does not perceive, the mathematics that underpins their work. It is more likely is that people vary in their degree of mathematics invisibility depending on what the individual perceives as mathematics, possibly compounded by his/her self-perception of being a mathematics-capable person or otherwise. Mathematics invisibility is a subjective experience, unique to the individual and able to be measured only in relation to something else (Keogh, Maguire, & O'Donoghue, 2010a). The intention to capture attitudes and beliefs around the utility of mathematics and its relevance to the workplace, ruled out any meaningful possibility of gathering quantitative or interval data. Instead, the participants were invited to indicate their agreement or otherwise with a range of Likert style statements, designed to capture their opinions (Likert, 1932). In a separate section, they were asked to indicate their frequency of use of mathematics across a spectrum ranging from 'Never or hardly ever' to 'about every day', on the basis that an empirical response could not be relied upon as being accurate or precise. This design determined that the resulting data set would be ordinal, presenting the challenge of how best to make sense of it.

It is common for researchers to assign numbers to response categories, in an ascending sequence, in order to convert Ordinal data into the more familiar Interval data, and to proceed to apply a broad range of parametric tests to establish validity and reliability and to draw conclusions from the reported regression and correlation statistics, e.g. Structural Equation Modelling. However, after careful consideration the authors decided that if one views the worker as a unique individual, influenced by unique combinations of a broad range of factors, behaving in numerate ways to solve problems and manage situations in workplaces which are themselves unique, then a generalisable, predictive model as may be produced by Structural Equation Modelling is not useful. An alternative complementary approach viz, Rasch Measurement Theory, was identified and used to provide reliability and validity to the data collected by the survey. A detailed explanation is available in the original thesis (Keogh, 2013).

That mathematics invisibility is particular to the individual, being unattainable, and always out of reach implies a tension between that which may be reported to have been perceived in comparison with that which is available to have been perceived (Coben, 2000b). Regardless of the individual's visibility of their use of mathematics in work, s/he may identify or locate, evaluate or interpret, act upon or communicate about mathematics containing situations or problems. By identifying the underpinning MKSC, it is possible to provide a platform for making them more visible. There are few options for highlighting the role of mathematics in work, especially in jobs where it is likely to be less visible and/or dismissed as something else. The most promising route was to use the existing frameworks in which all subject disciplines are calibrated for level, depth and range as is provided by the existing National Standards Frameworks.

Many countries throughout the world have a framework of qualifications under development or already in place. The Irish National Qualifications Framework (NFQ) provides for 10 levels, ranging from the Elementary or Foundation Level 1, rising incrementally through Levels 4 and 5, equivalent to the completion of

Table 3.1. Referencing the NFQ in Ireland to the EQF

NFQ Ireland	*Award Type*	*EQF*
1	Level 1 Certificate	1
2	Level 2 Certificate	1
3	Level 3 Certificate – Junior Certificate	2
4	Level 4 Certificate – Leaving Certificate	3
5	Level 5 Certificate – Leaving Certificate	4
6	Advanced Certificate – Higher Certificate	5
7	Ordinary Bachelor Degree	6
8	Honours Bachelor Degree – Higher Diploma	6
9	Masters Degree – Postgraduate Diploma	7
10	Doctoral Degree, Higher Doctorate	8

upper- secondary school in Ireland. This is followed by Level 6 indicating a post upper second level award often provided as a precursor to university entry. Levels 6 through to Doctorate Level 10 are the concern of the Higher Education System. For the benefit of the non-Irish audience, the NFQ is mapped on to the European Qualifications Framework (NQAI, 2009) (Table 3.1).

In Ireland, each of the 10 levels is classified by major strands of which there are 3 viz., Knowledge, Know-how and Skill, and Competence. Competence is a complex concept (Fitzsimons, 2002). It is sometimes used interchangeably with competency i.e. as an indicator of skills performance and is task oriented. The NFQ makes provision for skills performance, but uses the term competence to address a wider, more holistic concept elaborated in terms of Context, Role, Insight and Learning to Learn.

WORKPLACE EXPLORATORY TOOLS

On first contact, a workplace may seem to be chaotic, replete with a particular use of language and populated by people absorbed in their work, challenging the uninitiated to make sense of the purposes and processes on view. However, it is crucial for the researcher to become promptly oriented and equipped, not only to ask clarifying questions, but to understand the meaning of the answers. The authors recognised that observation alone would be unlikely to capture the activity and its cognitive underpinnings in the fullest sense. At the same time, becoming involved in the workplace risked compromising the research by inadvertent contamination. After careful consideration, the authors decided to employ a number of complementary tools and strategies (Table 3.2).

In addition to the many factors that contribute to mathematics invisibility, this present work added consideration of the way people talked about their work while

Table 3.2. Workplace exploration tools, aims and objectives

Tool	*Function*
Business Activity Model (Appendix 2)	Describe and depict the flow of work and information between workplace activities in an organisation from order initiation to completion
Work Practice Model (Appendix 3)	Establish and illustrate a fine grained account of a job and its surrounding context detailing constraints, conditions and options
Detailed Task Analysis (Appendix 4)	Illuminate specific tasks and their underlying mathematics knowledge, skills and competence as to mathematical provenance and calibrate their degree of complicatedness with reference to NFQ levels
Workplace Context-complexity Protocol (Appendix 5)	Capture the complexity of the workplace context that shapes the sophistication of the setting in which mathematics knowledge skills and competence are realised.

engaged in their work. Indeed, the researcher overheard dialogue between two work-colleagues that contained so many coded references as to render it beyond the comprehension of the 'outsider', despite English being their shared and commonly accented, first language. Similarly, verbal descriptions of the workplace alone, risked the omission of the depth of meaning as was captured by 'job-shadowing', i.e., accompanying the worker in the course of his/her work, and taking the opportunity to record the surrounding discourse, Chapter 5. The use of the Workplace Exploratory Tools was therefore accompanied by a comprehensive Workplace Transcript Discourse Analysis.

JOB-SHADOWING CASE STUDIES

It is customary for individuals to go about their work in apparently low-skills level, low-paid jobs, as a matter of routine, repeating a sequence of familiar actions, underpinned by operations that are performed automatically and selected to meet the specific needs of specific events. They may feature problem solving, decision making and situation management at a level commensurate with their formal role and performed as unspoken responses to familiar and unfamiliar situations. For this reason, it was necessary to ask questions such as "what happened", "how do you know", "how did you learn that", "what happens next", "what if", "what can go wrong" and so on, so that the participant could externalise elements of their work practice. This was captured on voice recording equipment and later transcribed and examined by the author using computer software, NVivo9, designed to support Qualitative Analysis (QSR International Pty Ltd., 2008). This offered the opportunity to grapple with the fabric of the context in which work was performed by thematic analysis. Although there were hundreds of instances of Numerate Behaviour detected in the transcripts, of more significance was the early indicators of a metric with which to capture the context of work, and which would subsequently inform the Workplace Contextualisation of Mathematics (Keogh, Maguire, & O'Donoghue, 2016).

IMPACT OF MATHEMATICS INVISIBILITY

While many factors are said to contribute to mathematics invisibility, to some extent at least, the authors argue that these affecters vary in intensity from one unique individual to the next, even between those sharing the same workplace. While the encounter between the worker and the work may be an entirely different experience due to personal background factors, the tasks, procedures, artefacts, and work practices are necessarily shared by people doing or collaborating in the same job.

It is beyond the current scope of this book to differentiate in any depth between competing theories of the nature of logical thinking, whether it is supported by whole integrated structures (Piaget & Inhelder, 1959), an inventory of schemata from which inferences may be drawn (O'Brien, 1998), a series of mental models of the real world (Johnson-Laird, 1983) or an innate, common sense ability (Colleran et al., 2003). However, it is apparent that mathematics concepts are not applied as discrete units of cognition in the initial stages of solving a problem or managing a situation. In this way, degrees of MKSC invisibility may indicate that mathematical ideas and ability have been successfully subsumed into an integrated cognitive resource with which people manage everyday situations. From this perspective, it appears that the continuous use of a set of knowledge and skills, in pursuit of a specific goal, may relegate the awareness of mathematics concepts in a person's work, but not necessarily the associated numerate behaviour.

Shining a metaphorical light on workplace activity has a two-tier effect. In the first instance, tasks that are performed out of habit or routine are brought to the fore by explaining their coherence and alignment with a string of objectives that are tuned to an overall objective. As a direct consequence, the underpinning knowledge, skills and competences that are 'hidden' in the shadows become more visible as rationale for the associated actions which are otherwise buried by shared assumptions and long-forgotten training. Completing the survey enabled the participants illuminate their own work practice much to their surprise in some cases.

CHAPTER 4

BROADENING THE SCOPE

Design and Key Findings of the National Survey of People at Work

An individual's Numerate Behaviour and attendant attitudes represent the culmination of formal, non-formal and informal learning, distilled by relevance, matured by experience and extended by insight, but may not be aligned with the propositional characterisation of mathematics associated with formal learning.

The performance of tasks underpinned by mathematics concepts and techniques, suggests that a person engages in what has been described as Numerate Behaviour (PIAAC Numeracy Expert Group, 2009), i.e. responding to mathematics information presented in different ways. By disaggregating what a person 'knows' and what s/he 'does', it is possible to establish links between the possible causes of mathematics invisibility and how that is expressed by the individual in the workplace.

The Programme for the International Assessment of Adults' Competencies defines Numerate Behaviour as

- *Responding to ...*
  - Identify, Locate or Access
  - Act upon or Use
  - Interpret, Evaluate/Analyse or Communicate
- *Information about mathematical ideas ...*
  - Quantity and Number
  - Dimension and Shape
  - Pattern, Relationship and Change
  - Data and Chance
- *Represented in a number of ways ...*
  - Concrete objects
  - Symbolic Notation
  - Images, Graphs, Maps and Visual Displays

... in order to solve a problem or manage a situation (PIAAC Numeracy Expert Group, 2009, p. 21).

The 'big ideas' of adult mathematics education in Ireland, are similar to PIAAC, characterised as:

- Quantity and Number
- Pattern and Relationship

DOI:10.1163/9789004381766_004

- Space and Shape
- Data handling and Chance
- Problem Solving (OECD, 2009).

The survey instrument at the heart of this work was designed to enquire about each Numerate Behaviour in response to each type of mathematics information (Domain) and sought to elicit responses regarding the awareness, frequency of use, and the implied visibility or prominence in the workplace. Furthermore, the structure and analysis of the survey items enabled the synthesis of individual numerate behaviours with each of the mathematics domains. The triangulation of measures of Numerate Behaviour with the surrounding discourse enabled the possible tensions to be associated with a workplace characterisation of mathematics.

In addition to completing the National Survey, a small number of workers were closely shadowed as they performed the tasks that underpin their job to enable the researchers develop a Detailed Task Analysis. By using the same instrument it was possible to report the *observed* frequency of Numerate Behaviour for comparison with the self-reported. The use of Differential Item Functioning (DIF) techniques, illustrated the contrast between the self-perceived and observed behaviours on the same scale and offered a profile and measurement of the extent of mathematics invisibility experienced by the case study participants. The case studies and findings are discussed in more detail in Chapter 5.

CHARACTERISATION OF WORKPLACE MATHEMATICS IN THE WORKPLACE

There are many personal factors that influence a person's development in unique combinations that may contribute to mathematics invisibility in particular. Similarly, there are organisational factors that both affect and are affected by each individual, and interact in complex ways that contribute to mathematics invisibility. Figure 4.1 outlines how the development of mathematics invisibility both at an individual and organisation level might occur.

A person is not born with an attitude or perception of mathematics, whether positive or negative. Rather, the perception of mathematics and its personal relevance, is developed over time and by persistent exposure to a variety of factors, and shaped by the norms and expectations of the environment in which it is first nurtured. The way in which a community communicates exerts a powerful influence and enables learning to take place. The absence, or otherwise, of a mathematics vocabulary neither excludes nor permits mathematics underpinned activity, only the way it is talked about and thought about. Over time, the collective experience develops a 'library' of historical situations that reinforce patterns of behaviour, to inform new challenges and the development of coping strategies. For want of a better explanation, the enabling knowledge, skills and competences are described as 'just common sense' i.e. anything but mathematics.

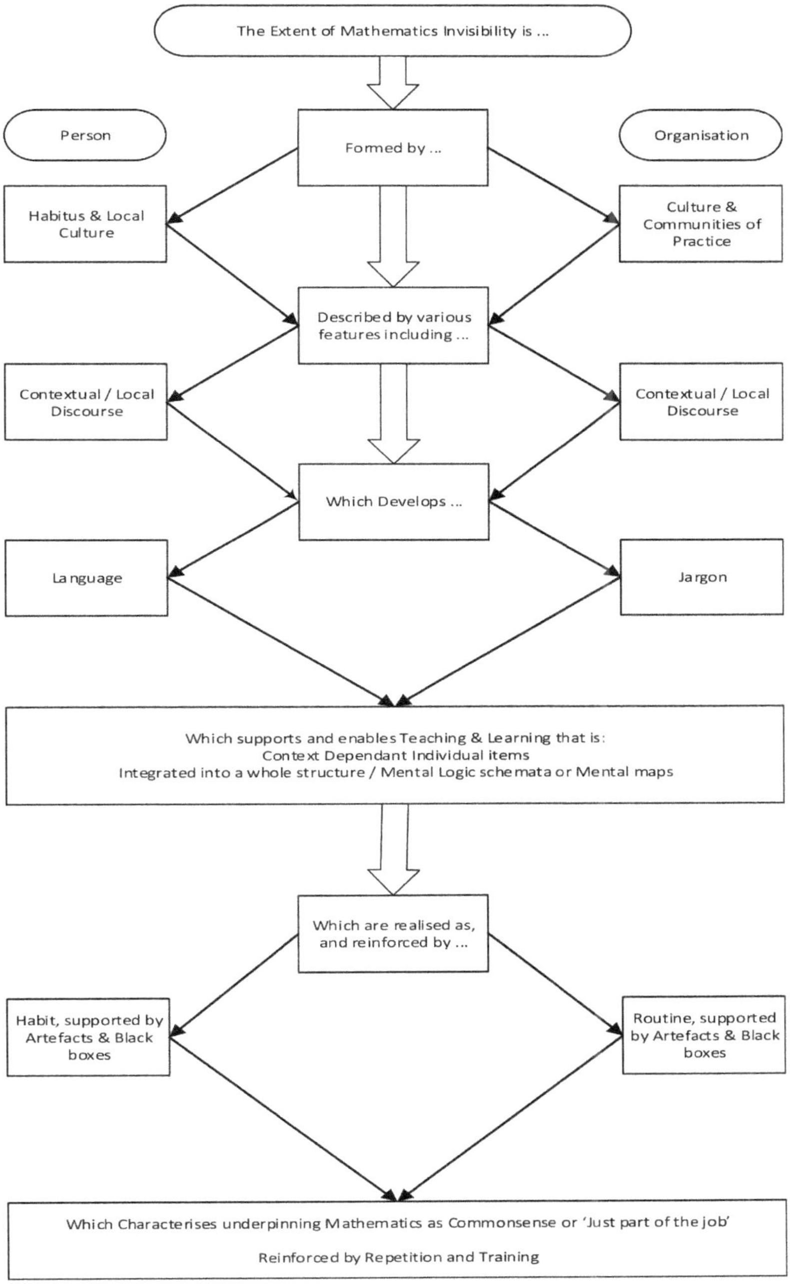

Figure 4.1. The paths to mathematics invisibility

The Workplace can be thought of as a collection of social activities that occur in a context made unique by its purpose, 'war stories', population and codified references. Instances of invisible mathematics in the workplace arise by a similar and parallel pedigree, traced by Figure 4.1.

While mathematics invisibility is not a directly observable phenomenon, it can be inferred from the factors that can be directly observed. Dealing with mathematics invisibility in the workplace requires the testing of the extent of awareness of mathematics and the frequency with which it is recognised. Awareness may be tempered by what one perceives to be 'mathematics' or not, and the extent to which it is disguised by process, procedure and discourse. Process and procedure highlights what the worker is required to 'do'. Workplace discourse describes and contains how workers talk about their work. Awareness and recognition of frequency of use reflect what the workers 'knows' about their work. The absence of such awareness implies that the frequency of use is irrelevant.

This survey was designed to tackle instances of mathematics invisibility from several viewpoints, viz., the awareness of mathematics described in several different ways, the frequency of use of mathematics described in different ways, and the extent to which process and procedure guided work-specific activity. Since the experience of mathematics invisibility is entirely subjective to the individual, the survey canvassed the attitudes and beliefs of people at work in Ireland.

DEMOGRAPHICS

The survey sample comprised people at work in Ireland, in a ratio of female respondents to males of 1.61:1. CSO data at the time of the survey indicate males outnumbered females at work by a ratio of 1.16:1 (CSO, 2010). This work does not intend to report on gender-related variance.

The survey enquired about education level and the highest level at which mathematics was studied by the respondent. These response categories were re-coded to align them with the National Framework of Qualifications provisions. The sample achieved (N = 449) was large in relation to the resources available to this research, and ranged across the full spectrum of mathematics education level, gender, and job skills level classified using the UK Standard Industrial Classification UKSOC 2010, in which jobs are classified into groups according to a concept of skills level and specialisation.

The extent to whether mathematics is invisible varies from one individual to the next and is a function of what is visible in comparison to what is available to be seen. At the outset, survey participants were invited to articulate the skills that were important to their job. From among 1200 replies, the term 'mathematics' or any other skill that is enabled by mathematics, e.g. accounting finance, budgeting etc., was cited only 4% of the time. Interestingly, Literacy was mentioned only 1% of the time (Figure 4.2).

Subsequent analysis of the data revealed considerable tension between the personal relevance of mathematics and the responses to enquiries about specific instances of the use of mathematics. The self-reported awareness of mathematics, albeit described in a variety of ways and applications, and the frequency of its use, stood in stark contrast to its almost complete absence from the list of important work related skills.

Figure 4.2. Statements of key job skills rendered to highlight the relative insignificance of 'maths' as an essential skill

It is clear that the 'maths', highlighted by the red arrow, barely features at less than 4% of the whole. It is as though MKSC does not come to mind, in the same way as Literacy and Reading skills registering at less than 1%, seemingly being taken for granted.

The findings of more detailed enquiry items, worded to reflect more realistic situations, depict quite a different narrative. It can be seen from Table 4.1, that the vast majority acknowledge very high levels of their awareness of the utility and relevance of mathematics in their workplaces, by Agreeing or Strongly Agreeing with the proposition of the survey item – with a few highlighted for special mention.

In sum, the data reveal that despite its apparent personal irrelevance as indicated by its absence from the list of important skills, the vast majority of the survey respondents routinely use and rely upon mathematics to do their work. Their initial attitude to

Table 4.1. Awareness of the role and utility of mathematics in work

Facet	*Survey Item Text*	*% Respondents Agree or Strongly Agree*
Examples of Numerate Behaviour	Quantities and measurements are a good basis for comparison between 2 things	88.3%
	Diagrams, plans, models, graphs and charts are useful when discussing work with colleagues	78.4%
	Differences in shapes and patterns between jobs are a useful way of comparing them	62.8%
	Graphs, charts, maps and schedules contain information I need for my job	77.4%
	The solution to a work problem is based on data and likelihood	69.8%
	Thinking about how a problem came about and about how effective the solution was, is good preparation for the next time a problem occurs	96.8%
	Knowing what to expect in work can give an early indication of a problem	90.9%
	Work problems can be solved with the help of graphs or timetables or maps or numbers	66.9%
Context Effect	It takes time to learn the lingo we use in work	76.4%
	I adapt the 'off the job' training I get for the job I actually do	86.2%
Explicit Reference to Mathematics	I use mathematical information in work	88.2%
	Maths play an important part in my work	68.8%
	The ability to calculate and estimate is important in my job	84.3%
	I pick out mathematical information to do my job	75.0%
	I pass on to another person any work problems that involve quantities or numbers	17.4%
	Mathematics is important in the world generally	94.4%
Process and Procedure	The work I do is straight forward	42.2%
	I follow the procedure to do my job	77.6%
	There is a procedure for dealing with many of the situations that arise in work	83.1%

mathematics is directly opposite to their attitude having completed questions that prompted them to account for their use of mathematics concepts and skills throughout the survey. So much so that it is clear that workers are very aware of the specific applications of mathematics in the service of identifying a problem, evaluating it, communicating about it and acting upon it, frequently. At the same time, this ubquity and reported utility of mathematics is not referenced as being important.

It is clear too that workplace activity is controlled by processes and procedures, to which the vast majority comply, to meet almost every anticipated situation. Typically, processes and procedures are codified by references that are context-specific, surrounded by a discourse developed for efficiency and meaning within that context alone. This shines a light on the work that must be done, rendering the underpinning knowledge skills and competence invisible. In other words, workers hardly acknowledge the role of mathematics in work while being very aware of it and using it frequently.

Procedures proliferate in workplaces to maintain standards, minimise scope for error and omission, and optimise service levels. Since the vast majority follow procedure to do their job, it seems a reasonable inference that work would be straight forward. This is not the experience reported by the majority of the workers. Furthermore, since 86% express the need to adapt off-site training to the needs of their job, and that 76.4% report a particular language use in the workplace indicates that the contextualisation of the workplace features something other than identifiable knowledge and skills, however they come to be known.

The majority (86%) Agree or Strongly Agree that off-the-job training has to be adapted to local conditions and surrounded by the language that is specific to the location of the work (76%). The clear implication is that workers hardly recognise the role of mathematics knowledge skills and competence that underpins their work, while at the same time, acknowledging its presence, utility and high frequency of use. In addition, it is the vast experience of the respondents that off-site training is adapted to the local conditions of the work context for which it is intended, i.e. shaped by the particular context in which it is applied.

FREQUENCY OF NUMERATE BEHAVIOUR

The Frequency segment of the survey is founded on the construct of Numerate Behaviour in response to different Mathematics Domains. The underpinning hypothesis is that, while enquiries about the use of mathematics in general terms expressed from an academic standpoint may be dismissed by workers as belonging in the mathematics classroom, enquiries shaped by the workplace activities may be more readily recognised and acknowledged. In other words, whether the individual rejects the relevance of mathematics or not, survey items that are set in workplace contexts, are more likely to capture better, instances of Numerate Behaviour in work and enable workers to recognise their occurrence and frequency. The corollary expected is that evidence of awareness of Numerate Behaviour should be corroborated by similar levels of frequency of use. While there may be evidence of other complementarities relating to gender, job skills, age-range and mathematics education, this present work did not explore them at this time.

The authors acknowledge that how often a person does something or uses a piece of skill in work can be cyclical. There are time-sensitive jobs that synthesise information from other sources, e.g., quarterly reports, forecasting and project review. These could occur daily, weekly or monthly. Yearly is in the realm of 'hardly ever' for the purposes of this survey, but could denote numerate behaviour nonetheless. In this way, instances of Numerate Behaviour that occur monthly, weekly or daily represent high frequency of use. Monthly activities are likely to be more demanding, requiring identification, synthesis, evaluation, interpretation, acting upon and communicating about. They may carry higher levels of risk, materiality, accountability, i.e. factors that seem to contribute to the complexity of the context rather than the difficulty of the MKSC involved. There is a possibility that to those for whom 'never or hardly ever' is a persistent and ubiquitous response, mathematics is truly invisible. This is balanced by the probability that since this survey elicited evidence of very high degrees of recognition of the frequency of mathematics use, and that the workers comprising the sample represent a very broad spectrum of endeavour, there may well be jobs substantially devoid of Numerate Behaviour, suggesting the need for further study.

The reported frequency of Numerate Behaviours in responding to different mathematics domains, is summarised in Table 4.2, on the basis of at least monthly, i.e. including weekly and daily.

Clearly, there are very high frequencies of all numerate behaviours in encounter with all mathematics domains, although Space and Shape is persistently last in the ranking. This is particularly curious in the physical, three-dimensional world which is described frequently by reference to size, volume, capacity, space, shape, dimension, weight, height, length and width.

While an account of the frequency of occurrence of numerate behaviours in broad terms is informative in itself, the structure of the survey data permitted an examination of each numerate behaviour in encounter with each mathematics domain. The following section describes the awareness of and frequency of use of, each mathematics domain in turn (Table 4.3).

Table 4.2. Matrix of numerate behaviour (NB) – monthly or more frequently – in response to specific mathematics domains (MD)

Numerate Behaviour	*Quantity and Number %*	*Pattern and Relationship %*	*Space and Shape %*	*Data handling and Chance %*
Identify	78	81.4	62.6	80.2
Evaluate	72.7	80.7	55.8	84.5
Act upon	75.9	64.1	52.9	71.5
Communicate about	74.7	71.5	62.0	76.4

Table 4.3. The awareness of respondents in relation to their numerate behaviour in the domain: quantity and number

Survey item text	*% Respondents Agree or Strongly Agree*
Quantities and measurements are a good basis for comparison between 2 things	88.3%
The ability to calculate and estimate is important in my job	84.3%
I pass on to another person any work problems that involve quantities or numbers	17.4%

QUANTITY AND NUMBER

This is strong evidence that survey respondents are very aware of their use of mathematics, while at the same time, indicating that problems involving Quantity and Number (Q&N) are faced by the majority in work and not avoided. Furthermore, respondents recognise a high frequency of the use of Quantity & Number, as evidenced by their numerate behaviours (Table 4.4).

The frequency of the reliance on mathematics-informed behaviours is consistently high. This is revealed by the language of the survey items, designed for the workplace setting with minimal mathematics specific language, further underlining the disconnection between the reported personal irrelevance of mathematics co-existing with its widespread use. That the importance of mathematics is concealed by discourse and context-dependent utility, speaks of the power of the workplace, or any habitual context to bind and contain, and is problematic for any expectation of useful transfer.

SHAPE AND SPACE & PATTERN AND RELATIONSHIP

The domains of Space and Shape (S&S) and Pattern and Relationship (P&R), while disambiguated for the needs of mathematics pedagogy, are not so easily discriminated by people in work. These terms are loaded with multiple, non-mathematics connotations, and are likely to index different meanings. The design of the survey instrument included equal amounts of items enquiring about awareness and frequency of each of the mathematics domains. However, the survey item validation process guided by the Rasch Measurement Theory rejected several of these awareness items as non-performing and providing unreliable data for this reason. This indicated that the respondents interpreted enquiries regarding Space and Shape, and Pattern and Relationship in different and inconsistent ways, when thinking about their awareness (Table 4.5). The Case Studies, which follow in Chapter 5, paid particular attention to Discourse to the extent that it referred to Space and Shape, and Pattern and Relationship in depth, to develop a more profound understanding of this workplace phenomenon.

Table 4.4. The frequency of different numerate behaviours (NB) associated with the domain quantity and number (Q)

Numerate Behaviour	*Survey Item Text*	*Frequency*	
		Yearly %	*At least Monthly (incl. Weekly) %*
Act Upon	I work with amounts and quantities to solve a problem	23.2	75.9
Evaluate	I use reasoning skills on the numbers aspect of a problem to try and find a solution	22.3	76.9
	I make work judgments around priority, urgency, importance and so on, on the basis of the numbers involved	19.1	80.4
	I check on the quantities aspect of a work situation after a solution has been implemented to gauge how well it has worked	30.9	65.3
	I figure out the affect that quantities have on my work	26.6	68.0
Communicate About	I discuss quantities associated with my job with work colleagues	21.0	78.5
	I describe some things in work using numbers and mathematical ideas	26.2	73.5
	I describe the solution to a problem using numbers and quantities	27.9	72.1
Identify	I get work information in the form of quantities and numbers	21.2	78.0
Workplace Specific Activity	I make estimates in my head about work	10.7	88.5

Table 4.5. The level of awareness of respondents in relation to their numerate behaviour in the domain of space and shape and pattern and relationships

Survey item text	*% Responses Agree or Strongly Agree*
Diagrams, plans, models, graphs and charts are useful when discussing work with colleagues	78.4%
Graphs, charts, maps and schedules contain information I need for my job	77.4%
Work problems can be solved with the help of graphs or timetables or maps or numbers	66.9%
Differences in shapes and patterns between jobs are a useful way of comparing them	62.8%

There is measurable incongruence between the perceived capacity of S&S and P&R to represent mathematics containing situations when compared with usefulness for solving problems. While still reporting high levels of awareness, it is somewhat counter-intuitive that the source of mathematics information is not equally perceived as a problem-solving resource, especially when taking account of the frequency of use of P&R (Table 4.6).

Table 4.6. The frequency of different numerate behaviours (NB) associated with the domain pattern and relationship 1: any numerate behaviour & pattern and relationship

Numerate Behaviour	*Survey Item Text*	*Frequency*	
		Yearly %	*At least Monthly (incl. Weekly) %*
Problem Solve	When trying to solve a problem, I take into account that a change in one part of the problem can have an affect on another	9.0	89.7
	I use my knowledge of how different parts of the system affect each other to find a solution to a problem	14.0	83.7
Act Upon	I convert work information from one measuring system to another	35.0	60.9
	I make estimates about the information contained in charts, graphs, diagrams, timetables and the like	31.9	67.3
Communicate About	I discuss work drawings, schedules or spreadsheets with colleagues to make sense of them	28.2	71.5
Identify	I find what I want by sorting information into sequence	17.3	81.4
Evaluate	I use my knowledge of the system to trace a fault back to the cause	12.8	84.7
	I notice there is a problem when some aspect of my work differs from what I would normally expect	17.3	80.7
	I detect problems in work when the data does not seem to fit with normal expectations	21.3	77.9
	I use the information I get from a variety of sources to make judgments about work	12.8	86.7
	I figure out the affect one quantity may have on another in work	22.1	77.6
	I detect problems in work when something does not seem to fit in with the usual pattern	18.3	80.9

All Numerate Behaviours are reported as engaging very frequently with Pattern and Relationship-containing situations. This indicates the codified nature of work in addition to expectations, anticipation, and planning capabilities built up by exposure to a particular workplace. However, repetition forms habits of mind that subsume and contain the underlying MKSC, rendering it invisible and out of reach despite its widespread use (Table 4.7).

Table 4.7. The frequency of different numerate behaviours (NB) associated with the domain space and shape

Numerate Behaviour	*Survey Item Text*	*Frequency*	
		Yearly %	*At least Monthly (incl. Weekly) %*
Communicate About	I use measurements to describe the solution to a work problem	28.7	70.1
Evaluate	I measure dimensions in work	43.6	49.1
	I notice a problem in work when the measurements and dimensions seem to be out of line with the usual	35.1	58.1
	I think about the shapes involved in a problem to find the solution	53.4	46.6
	I read charts, graphs, sketches and the like to get information about my work	27.4	72.3
	I use reasoning skills on the shapes involved in a work problem when trying to figure out a solution	43.5	49.1
	I notice there is a problem in work when the shape of some part of it is different from usual	43.1	52.2
	I use my knowledge of shapes when trying to find a solution to a work problem	51.6	41.9
	I use reasoning skills on the measurements involved in a problem to figure out a solution	15.8	80.3
Identify	I get work information in the form of measurements and proportions	36.6	62.6
Act Upon	I use the shapes involved in a problem when trying to find a solution	47.9	44.7
Work-related activity	I use diagrams, sketches, models and the like when I communicate about work	39.0	61.0
	I measure things	30.8	68.4

While the prominence of measure is apparent, the frequency of engaging with Space and Shape is less than with other domains including Pattern and Relationship. These observations combine to suggest that Space and Shape are less useful and less perceptible in underpinning workplace activity. This is counter-intuitive in a world that is described in three-dimensional terms. This is further evidence of the role of Space and Shape being diminished due to its ubiquity and familiarity.

DATA HANDLING AND CHANCE

Further tensions were observed in the domain of Data handling and Chance, regarding reported level of awareness (Table 4.8).

Table 4.8. The awareness of respondents of their numerate behaviour in the domain: data handling and chance

Survey item text	*% Respondents Agree or Strongly Agree*
The solution to a work problem is based on data and likelihood	69.8%
Work problems can be solved with the help of graphs or timetables or maps or numbers	66.9%
Thinking about how a problem came about and about how effective the solution was, is good preparation for the next time a problem occurs	96.8%

There is strong evidence that workers use prior knowledge to anticipate and resolve problems (96.8%). Nevertheless, workers only recognise their use of the data that describe the problem when seeking a solution to a much lessor extent (c. 67%).

That there is substantially less recognition of the problem-solving capacity of the available data than the capacity to extrapolate from experience, suggests that there is a disconnection between the representation of data and the imagining of it. This may echo the 'common sense' perception of work skills, once having been learnt, becoming subsumed into an automatic operation comprising identifying, evaluating and acting upon mathematics containing situations.

The reported frequency of the use of data is quite high across all numerate behaviours, especially interpreting information, which is at odds with the reported levels of awareness. This is another strong statement of the reach of the workplace-context effect, i.e. that workplace learning is confined and contained by the context in which it was learned, because it is made sensible by the surrounding conditions and less visible otherwise.

So far, the lens of enquiry has been shaped by the construct of numerate behaviour. The survey also enquired about training related behaviours. Typically, a worker is trained to cater for situations (Table 4.9) which may be anticipated from past experience, by providing a palette of 'appropriate solutions'. However, the challenge of matching the optimal solution to the perceived problem persists.

Several of the survey items mentioned work explicitly to set the enquiry in the workplace context by drawing attention to workplace-specific activities (Table 4.10).

Two facets emerged from these survey items. In the first instance there was widespread agreement that in every job workers apply ready-made solutions to problems that are expected to emerge from time to time, perhaps as a result of their recurrence over time. That the worker makes a choice of the optimum solution from a range of possible solutions indicates that it is not a simple matter of matching the solution to a problem, but rather that the problems arise in ways that are nuanced by a variety of factors. This echoes the finding that despite the controlling influence of process and procedure, the majority of workers do not find their work to be straight forward.

The design of the survey items included the sub-categories that enabled the data to be reclassified as either relating to Problem Solving or Managing Mathematics containing Situations behaviours, in encounter with all mathematics domains.

Problem solving behaviours are exemplified by keeping track, solving, selecting, identifying, evaluating, whereas Managing Mathematics containing situations

Table 4.9. The frequency of different numerate behaviours (NB) associated with the domain data handling and chance

Numerate Behaviour	*Survey Item Text*	*Frequency*	
		Yearly %	*At least Monthly (incl. Weekly) %*
	I interpret information from a variety of sources to form an opinion about a work situation	8.6	90.4
Evaluate	I keep track of potential problems by comparing actual with expected outcomes	20.7	78.5
Work-related activity	From among the possible solutions, I select the one most likely to solve the immediate problem	13.3	84.2
	I merge work data from different sources to create an overall report	26.2	73.3
Identify	I solve a problem by exploring the data for a likely explanation	21.7	77.3
Communicate About	I describe a work situation using the data involved	22.8	76.4

Table 4.10. Frequency of adapting training to work-specific contexts

Domain	*Text*	*Frequency*	
		Yearly or less	*At least Monthly (Includes Weekly)%*
Training	I apply ready-made solutions to familiar problems that arise in work	13.5	85.7
	From among the possible solutions, I select the one most likely to solve the immediate problem.	13.3	84.2
	When trying to solve a problem, I take into account that a change in one part of the problem can have an affect on another	9.0	89.7
	I use my knowledge of how different parts of the system affect each other to find a solution to a problem	14.0	83.7

denote reasoning, gathering, merging, enabling, communicating, noticing, and using. The survey items were re-classified by reference to whether they were deployed to solve problems or to manage mathematics containing situations. Using this categorisation across all mathematics domains, provides the following profile. While all Numerate Behaviours are reported as engaging with Space and Shape, the regularity at which this occurs is much reduced in comparison with the other domains (Table 4.11).

The survey data report that, at least monthly, around two thirds of the respondents manage mathematics containing situations (Mng) in the domain of Space and Shape yet less than half recognise its role of Space and Shape in solving problems in work. It is noteworthy that the respondents discriminated against the term 'dimension', the activity of measuring 'things' being recognised 40% more frequently than 'dimensions', which suggests a particular and negative connotation borne by mathematics terms.

These data combine to suggest that Space and Shape are less useful or less perceptible in underpinning workplace activity, which is in keeping with the levels of awareness. The utility of Pattern and Relationship is more readily recognised (Table 4.12).

All Numerate Behaviours are reported as engaging very frequently with Pattern and Relationship containing situations. The reclassification of the items as either problem solving or situation management shows that these behaviours are more associated with Problem Solving (P 84%) than in Managing (M 73%). Although there is a persistent 20% in the Yearly or less category, it may be overstated on the basis of items that score well above the mean.

Table 4.11. Frequency of problem solving and managing mathematics containing situations using space and shape

Survey Item Text	*PS or Mng*	*< = Yr %*	*> = Mt %*
I notice a problem in work when the measurements and dimensions seem to be out of line with the usual	PS	35.1	58.1
I think about the shapes involved in a problem to find the solution	PS	53.4	46.6
I notice there is a problem in work when the shape of some part of it is different from usual	PS	43.1	52.2
I use the shapes involved in a problem when trying to find a solution	PS	47.9	44.7
I use reasoning skills on the shapes involved in a work problem when trying to figure out a solution	PS	43.5	49.1
I use my knowledge of shapes when trying to find a solution to a work problem	PS	51.6	41.9
Problem Solving activities Mean %		*45.8*	*48.8*
I use measurements to describe the solution to a work problem	Mng	28.7	70.1
I read charts, graphs, sketches and the like to get information about my work	Mng	27.4	72.3
I get work information in the form of measurements and proportions	Mng	36.6	62.6
I use diagrams, sketches, models and the like when I communicate about work	Mng	39.0	61.0
I measure dimensions in work	Mng	43.6	49.1
I measure things	Mng	30.8	68.4
Managing mathematics containing situations Mean %		*34.4*	*63.9*

The contrast between the perceived usefulness of Space and Shape in comparison with Pattern and Relationship is striking, as is the reported frequency of the use of Data handling and Chance (Table 4.13).

All Numerate Behaviours are reported as being used at least monthly in either Problem Solving or Managing mathematics containing situations (80%) in the case of Dh&C, whereas around 20% do not. Problem solving involved Acting Upon, Identifying and Evaluating only, whereas Managing involved Acting upon, Identifying, Evaluating and Communicating about. However, there is a substantial proportion of respondents who said that they do not use the data to trace a problem; they do not

Table 4.12. Frequency of problem solving and managing mathematics containing situations using pattern and relationship

Survey Item Text		*Frequency*	
		Yearly %	*At least Monthly (incl. Weekly) %*
When trying to solve a problem, I take into account that a change in one part of the problem can have an affect on another	PS	9.0	89.7
I use my knowledge of how different parts of the system affect each other to find a solution to a problem	PS	14.0	83.7
I find what I want by sorting information into sequence	PS	17.3	81.4
I use my knowledge of the system to trace a fault back to the cause	PS	12.8	84.7
I detect problems in work when something does not seem to fit in with the usual pattern	PS	18.3	80.9
Mean %	*PS*	*14.3*	*84.1*
I convert work information from one measuring system to another	Mng	35.0	60.9
I make estimates about the information contained in charts, graphs, diagrams, timetables and the like	Mng	31.9	67.3
I discuss work drawings, schedules or spreadsheets with colleagues to make sense of them	Mng	28.2	71.5
I notice there is a problem when some aspect of my work differs from what I would normally expect	Mng	17.3	80.7
I detect problems in work when the data does not seem to fit with normal expectations	Mng	21.3	77.9
I use the information I get from a variety of sources to make judgments about work	Mng	12.8	86.7
I figure out the affect one quantity may have on another in work	Mng	22.1	77.6
Mean %		*24.1*	*73.2*

Table 4.13. Frequency of problem solving and managing mathematics containing situations using data handling and chance

Survey Item text		*Frequency*	
		Yearly or less	*Monthly or more*
I keep track of potential problems by comparing actual with expected outcomes	PS	20.7	78.5
From among the possible solutions, I select the one most likely to solve the immediate problem	PS	13.3	84.2
I solve a problem by exploring the data for a likely explanation	PS	21.7	77.3
Mean %	*PS*	*18.6*	*80.0*
I describe a work situation using the data involved	Mng	22.8	76.4
I interpret information from a variety of sources to form an opinion about a work situation	Mng	8.6	90.4
I merge work data from different sources to create an overall report	Mng	26.2	73.3
Mean %	*Mng*	*19.2*	*80.0*

synthesise data from different sources; or describe work in terms of the data involved, or compare the actual outcome with the expected outcome. At the same time, the vast majority say that they collect and interpret data from a variety of sources.

The survey respondents presented and reinforced the paradox that workers are very aware of their numerate behaviour, and its prominence in the performance of their work. At the same time mathematics is afforded almost no importance from their perspective.

SUMMARY

The authors generated insights from the survey data related to the following research issues:

It is clear that there exists a tension between what workers perceive as key skills and what they report as the role of mathematics in their work. This is especially stark when associated with the term 'mathematics', yet when invited to reflect on their activities in work, there are high levels of awareness of, and high degrees of frequency of the use of, mathematics, albeit described in a variety of context-sensitive ways.

1. *Workers do not report mathematics as a skill that is key to their work.*

There is strong evidence, detected from a number of perspectives regarding the encounter of a range of numerate behaviours and mathematics domains, that workers are very well aware of the numerate behaviour when it is set in their familiar surroundings and couched in non-mathematical language. Furthermore, they report very high frequency of use of knowledge, skills and competence underpinned by mathematics.

2. *Workers are well aware of their numerate behaviour in encounter with a range of mathematics domains.*
3. *Workers recognise the frequency with which they use and rely on a range of mathematics domains, albeit described in different ways.*

Workers use all numerate behaviours in encounter with all mathematics domains to solve work-based problems and to manage mathematics containing situations

4. *Workers behave in numerate ways to solve problems and manage mathematics containing situations*

Modern workplaces are complex configurations of processes and procedures that are designed to choreograph the application of disparate skills sets and expertise embodied in different, yet aligned, operational units, departments and communities of practice. The challenge is to ensure the smooth interaction of these units with optimum efficiency to 'get it right every time', to maintain standards and minimise waste in time and materials. There are standard operating procedures for almost every situation, with which workers comply most of the time, but it is not as simple as that.

5. *While work is shaped and contained by processes and procedures, to which they predominantly comply, they do not perceive their work to be straightforward*

The four mathematics domains that framed this work were regarded with almost equal levels of awareness and used with similar frequency, with the exception of Space and Shape. In every instance the recognition of shape and space was much less that the other three.

6. *The notion of space and shape, while a prominent descriptor of the physical world, is not acknowledged as a factor of the workplace to the same extend as the other mathematics domains*

Typical features of recruitment policy include the requirement for formal and relevant qualifications, accompanied by the expectation of relevant experience measured in years. At the same time, workers report that the frequently have to adapt off-site learning to their work context, in addition to 'learn the lingo' and 'how we do things around here'. This is underlined by the power of context to bind and constrain knowledge, skills and competence, making transferability problematic.

7. *While employers value competence gained tacitly on-the-job, they privilege formal accredited learning in recruitment, is highlighted, yet, having selected suitably experienced candidates, they proceed to re-train them in 'how we do things around here'.*

Despite widespread awareness of-, and very high frequency of the use of mathematics-underpinned knowledge skills and competences, they are not deemed to be personally or professionally important. The spotlight on what a worker 'does' surrounded by the 'lingo' of the context, in addition to the capacity for mathematics being in the service of many intersecting contexts, tends to render it invisible.

8. *The workplace context effect is that the mathematics-saturation of the workplace context can be disguised in many ways perhaps confirming their self-perception of 'not being a maths person'.*

The survey data were provided by people of varying mathematics education backgrounds, who were engaged in a broad spectrum of work. Having exhausted the potential of the traditional methods of reporting proportionality and item response frequency, the authors sought validation for the breadth of the findings by adding depth to the enquiry, by conducting several case studies of work for which mathematics is neither in evidence, *prima facie*, nor a qualifying component. These case studies and their significance are reported in Chapter 5.

CHAPTER 5

JOB SHADOWING

Four Case Studies of People Engaged in Their Job

JOB SHADOWING PEOPLE AT WORK

The well documented putative factors contributing to mathematics invisibility in the workplace and the existence of overt mathematical activity in a company, say, in engineering or design departments, sets up the dichotomy of 'maths' and 'non-mathematics' departments. In other words, where mathematics is easily visible and associated with a particular department or type of job, all other activity is reinforced as involving common sense or anything but mathematics. However, as mathematics underpins different contextualisations of work, and may be invisible to a greater or lesser extent depending on the individual, necessarily directs the research to contexts where there is little or no *prima facie* evidence of mathematics activity. A number of job shadowing case studies across two companies were completed to provide mutual corroboration in the general sense and underlined nuances that were unique to each setting, serving to highlight the possibility of identifying common components yet context-specific detail.

Case-study candidates were selected on the basis that there was little or no superficial evidence of mathematics activity. However, identifying and setting boundaries to such jobs to be 'job-shadowed' required the application of a number of devices developed by the authors to peel away a succession of organisational and procedural layers to reveal the MKSC at its foundations. In the first instance, the companies hosting the case studies were selected for their willingness and the generosity of their freedom of access.

There are four case studies at the heart of this present work, two of which were hosted by Company A, and two by Company B. This segment describes and models each company and the case studies they hosted at increasing levels of magnification, to identify the MKSC that underpins each job. The alignment of the identified MKSC with the NFQ is explained, demonstrated and illustrated in each case. The challenges of dealing with instances of Mathematics Invisibility in the workplace are discussed and a potentially fruitful approach suggested.

In every case study reported the worker was invited to take part in the National Survey of People at Work. The timing of the introduction of the survey instrument to the worker was a difficult balance to strike. The purpose of the survey was to provide a backdrop against which the individual's self-reported Numerate Behaviour could

 | DOI:10.1163/9789004381766_005

be examined and compared with that of his/her behaviour as observed in the case study, and interpreted with the support of the surrounding discourse.

CASE STUDY HOST COMPANY A

Commercial Purpose

Company A employs 200 people to provide document management services on the island of Ireland. It is a very competitive business, operating in the midst of economic turmoil in a contracting market, with a widespread fear of job-loss and long-term unemployment. Although the 'paperless office' is beginning to be realised, industry still produces very large quantities of text on paper which are legally required to be retained for 6 years. In many cases, managing the volume of documents constitutes a significant business overhead in terms of administration time and office space. In addition, sensitive documents, property deeds and computer back-up media are often required to be stored in a secure, fire-proof location off-site, as a condition imposed by insurance providers. Company A's customers can choose to have their documents and/or computer media, collected from their own premises, catalogued individually or in boxes, and stored securely until recalled, or redelivered in accordance with an agreed schedule. Each individual item is identified by a unique barcode to enable an internal tracking system. Every time the location of an item is changed, it must be logged by a handheld scanning device. In this way, the computer system 'knows' where each piece is stored and its movement history. New items for storage are tagged with a barcode at the point of collection and added to the inventory on arrival at the storage facility. Error conditions arise when an item is not found in the location indicated by the computer system or when the movement logged is not consistent with its history. These are recorded as exceptional items and are the subject of a separate 'audit' activity.

The site visited for this research had a storage capacity exceeding 100,000 cubic metres, sufficient for 3 million items, files and documents, spread across two adjoining but differently designed buildings, comprising up to 4 floors, each containing rows of storage racks, several tiers high and several rows deep. There are secure vaults with controlled access, specially designed to contain property deeds, legal briefs and computer media. The customer base ranges across the spectrum of professions and industries, public services and a small but growing number of private individuals. Many customers have requirements conditioned by day of the week, e.g., to recycle computer backup media, but storage and retrieval transactions differ from day to day, in reaction to customer service requests. The underlying service assurance is that requests for documents received before noon on day 1, will be delivered no later than noon on day 2. Requests for items that are urgently required by the customer to be serviced within the hour, is an optional, premium service and is priced accordingly. Customers may specify that items in storage after 6 years have elapsed are to be destroyed without further notice.

The first step was to deploy one of the tools developed to capture the setting in which work took place, i.e., to model the activity of the business and to enable the researcher to ask better questions and to understand the answers. After a brief meeting with the chairman to re-affirm agreement on terms and conditions of access, the researcher was brought directly to the Document Storage Operations Centre and introduced to the manager. He explained the functions and interactions of the different activities in sufficient detail to enable the preparation of a first draft of the Business Activity Model (BAM), subsequently confirmed and refined, Appendix 2 exemplar. A completed BAM provides a high level profile of the full landscape of the activities of the business, their sequence and their interdependencies The main purpose of the BAM is to chart the context in which particular groups of tasks are performed including the source of their inputs and the aim of their outputs, giving them sense and purpose. Guided by the draft BAM, the researcher traced different facets of the operations in conversation with the staff, asking questions for clarification and making corrections to the BAM as necessary, in preparation for selecting jobs as suitable candidates for case study.

Business Activity Model – Company A

The company structure comprises seven centres of business activity, as detailed in the BAM, Appendix 2. All incoming telephone contacts are directed to Customer Service. Calls concerning potentially new customers or new requirements for existing customers are directed to Sales, where, in conjunction with technical input from Administration, a Service Level Agreement (SLA) is offered to the customer, detailing terms, conditions and price. A copy of the SLA is stored on an internal computer network server. There may be several cycles of contact between Sales/Admin and the customer in order to reach a mutually beneficial agreement. Service Requests are also fielded by Customer Service personnel who, having checked the customer's credit status, produce a detailed Work Order, which is recorded in the Work Order Log and stored on the internal computer network. Work Orders, whether instructions to collect and re-file or to retrieve and deliver, or both, are attended to by the Warehousing and Transport Activities. Collections of previously stored items are logged as being 'Returned' by Warehousing and subsequently re-filed. Collections of new items are catalogued and logged on the computer tracking system before being stored in the warehouse. Warehouse 'Pickers' retrieve and re-file items as detailed in the Work Order. Collections and deliveries are carried out by van/truck drivers, each responsible for a geographical area or Route, although these may overlap at the discretion of the Transport Manager, who works in close cooperation with the Warehouse Team Leader. Billing is an automated function fed by the terms of the SLAs and completed Work Orders.

The authors did not consider Accounting, Payroll, Statistics, Treasury/Finance or the Transport Management activities as potential case studies, because of their more obvious associations with mathematics concepts, and especially since more interesting candidates were available, i.e. 'Warehouse Picker' and 'Van Driver'.

Case Study 1 – Warehouse Picker 'Rob'

The role of the Warehouse Picker was deemed to suit the requirements of the research ideally. It appeared to be a physically demanding job, involving lifting and carrying heavy boxes, climbing stairs and ladders, frequently having to move several boxes to reach a particular box. It bore a relatively low status level and did not require any particular academic qualifications. Each Picker worked alone, usually, receiving instructions in batches from the team leader, and preparing deliveries for a delivery driver whom he did not need to contact in the normal course of his work. There were no obvious signs of mathematics activity. 'Rob' is a utility warehouse worker, at times picking, re-filing and making space available. His suitability for the study was based on the author's impression that his approach to his work would be more deliberate, and more readily observed than someone whose routine was performed at the level of automatic operations, and, perhaps unavailable or less available for enquiry.

Contextual Factors

Observing work in practise reveals the physical engagement of a person in a purpose which may or may not be immediately discernible. The Workplace Practice Model (WPM) tool was designed by the authors to discover the nature and extent of the rationale, objectives, constraints and conditions that guided Rob's work and to make sense of its interconnection with the surrounding business activities, Appendix 3 as exemplar. The WPM enables the researcher to zoom in on selected components of Rob's work and to specify it in finer detail. It brings to the surface the range of relevant inputs and the decisions and actions required to produce the expected outputs, taking account of conditions, constraints and priorities, and the circumstances under which they change. In this case, the WPM focuses on the actives of the warehouse 'Picker'.

The main function of the Picker is to retrieve a number of specific items from their locations in the warehouse(s), and make them ready for delivery to customers in accordance with Work Orders. The Warehouse Team Leader interrogates the Work Order Log, extracts the orders appropriate for the shift and allocates them to individual Pickers. Knowledge of the type of the customers and their patterns of service orders permits the work to be distributed fairly, taking the customers' needs and the capability of the Picker into account. The Picker is given a sheaf of customers' printed Work Orders, detailing the type and barcode of the items to be retrieved. He is equipped with a handheld scanning device with which to download a 'pick list', i.e., the barcodes and warehouse location codes of the requested items. The handheld software provides search and sort options, in addition to verifying the identity and ownership of a barcoded item. It is also used as a data recording device to log changes in status e.g., from being stored in a particular location to having been retrieved in preparation for delivery.

It is usual for the picker to have to pick 60 items within an hour. While this would seem to represent 1 item per minute, several picks can be clustered together and easily retrieved. In addition to fetching items from storage, the picker must reconcile the items picked with the work order for each customer and group them for their respective delivery route. Each item bears an 8 digit, item ID barcode. There may be a second ID code to indicate a specific 'banker box' (cardboard file storage box), in which a file is stored. Other codes may be visible on the item having been placed there by the customer or by a rival storage company, having previously been stored by them. The item tracking software permits the customer-supplied code to co-exist with that allocated by the company. In this way the internal tracking system links the item ID Code with its last reported location, including the ID of the box in which it is filed if appropriate. The occasional presence of rival company barcodes on the item can be a source of mis-identification and error.

Environment

Fetching specific items from their locations is not helped by the layout of the storage facilities within the warehouse complex. The storage capacity of the warehouse evolved in response to rapidly growing demand rather than in line with a coherent location plan. The storage racks were fitted wherever space became available and labelled with location codes on an *ad hoc* basis and without consulting the Warehouse Team Leader. Secure vaults were added, as were specialised storage racks for different kinds of computer media. The warehouse was extended into an adjacent building to meet the demand for more space, adding to the incoherence of the location referencing system. By this stage of the company's expansion, it had become impossible to reorganise the warehouse layout and to amend the transaction history without shutting down the facility for a prolonged period. The result is a location coding system that is not consistent in every building, and every floor, that does not take into account the relative distance between locations on the same or an adjacent floor, the proximity of vaults, cargo elevator and stair wells, and the weight of the item to be picked. The location code comprises 10 alphanumeric digits e.g., LEV118J10C to indicate Building (LEV), and floor (1), row (18), bay (J), box position number (10) and shelf (C). The pick list, downloaded to the picker's handheld device, directs the picker to retrieve the next item by reference to the alpha-numerical sort sequence, not necessarily the next-nearest item. For example, the typical warehouse floor has 52 rows of racks, numbered 01 to 52, each divided into 24 bays, labelled from A to X. There is a wide aisle at bays E and K, permitting access for a hand-truck to adjacent racks without having to pass through at either end. There are 3 shelves in each bay, A, B and C, rising from floor to ceiling, each containing up to 30 boxes. Access to shelf C requires the use of a ladder. The picker has to notice this before entering an aisle so that he can bring the ladder with him, rather than having to retrace his steps. Only the most highly active areas of the warehouse are permanently lit by electricity. Pickers often use the light emitted by the handheld as a torch to guide them to where they might need only one light, and not the inconvenience of switching off many lights *en-route*.

Since these circumstances directly affect the pickers, each has developed his (they are all male) own efficiencies so that he is not crossing and re-crossing the warehouse floor, unnecessarily. He looks ahead to the next items on the pick list to ensure that he visits an aisle once only. If a ladder has been used, he must replace it to where it will be visible to the next picker. He may also use one of several menu-driven, look-up options to re-sort the handheld pick list by reference to the customer in order to ensure completeness. The scanning function of the handheld device enables the picker to verify that the correct item is being retrieved and to record its new location in order to update the item location-history log. The handheld device maintains a count-down clock to remind the picker of the volume of picks remaining, which may have the sub-conscious effect of causing him to hurry. There may be a map of the warehouse, but it could not be found, and is not used either in training or work. Untrained persons are not allowed in the warehouse unaccompanied. One is expected to learn the warehouse layout by trial and error. Training is on-the-job and takes the form of progressing from accompanying an experienced Picker, to having sole responsibility for short, and then full lists, under the pressure of time.

Job Shadowing – Case Study 1, 'Rob'

Having received his printed customer Work Orders, 'Rob' downloaded his pick list by placing his handheld scanner in a cradle which is directly connected to the computer network. There he manipulated menu options to execute the download procedure. He could not use the wireless option, because his device was not working properly as he had dropped it previously. The downloaded pick list was sequenced by location code in alphabetical order, together with the item ID code, and that of its container. Rob stated that the handheld "does everything except the pick". From this standpoint, the Picker's job would seem to be straight-forward. Rob expressed the view that if the handheld could do the pick, there would be no need for him.

In the first instance, Rob studied the handheld pick list, occasionally consulting the Work Orders, for an overall view of the pick. While the sequence of the pick list is a helpful guide, it does not cater for the design flaws of the warehouse facility and its labelling system as previously described. The Work Orders also informed him about the van drivers he is preparing the deliveries for, and what their preferences are. He noted the combination of buildings, floors, vaults, boxes, cartons, files, deeds, tapes and so on, that was involved in his workload and planned the sequence of his work accordingly.

Rob did not follow the handheld pick list rigidly. For example, he noticed that some of the requested items had been recently returned, soon to be re-filed, and as such may be located in a particular and convenient area, i.e., the 're-file' table, which he checked first. This was more likely to achieve a positive result due to recent surrounding circumstances in which the re-filing personnel had been temporarily re-assigned to other duties, allowing the volume of returned files to build up. Then he made his way to the next nearest location by making the most efficient use of the access routes, location of stairs and gaps in the storage racks, displaying an impressive degree of spatial awareness in multiple dimensions.

The interpretation of a simple monomial such as the algebraic location code, in the context of the warehouse and its physical constraints, guide Rob to the storage bay and shelf, but no more. To find the specific item, Rob engaged in an activity that was quite different from finding a uniquely identified storage bay. Depending on the rack design, a shelf could contain up to 30 boxes, arranged in 9s or 10s, two or three deep, at floor, waist or head height, such that only the front 9 (or 10) boxes were visible to the scanner. Rob was faced with two options; he could scan each barcode until he heard the single toned 'beep' signalling the correct box, or he could search manually. Either option involved moving boxes to access the next tier, which was particularly hazardous when searching on the top shelf. The scanning option was likely to return the multi-toned 'error' sound, but, as this could be caused by scanning a rival company's barcode, it would need to be re-scanned. Rob chose instead to memorise the barcode and search for it, reasoning that although there are 8 digits in the barcode, he need only remember the 4 lowest order values, because the "whole barcode would be the same, except for the last 4 digits". This search routine is unavoidable, given the imprecision of the location code. However, the necessity to move boxes can create an error condition – if one or many boxes, having been removed to find a hidden box, are incorrectly replaced. The tracking system will continue to report the original location until a 'new' movement is logged.

While the computerised tracking system is fool-proof, human error can disrupt the expected flow of work. When, on occasion, an item was not found at its recorded location, Rob adopted a search strategy, by applying his knowledge of the system and his experience of how previously mislaid items had been found. He checked adjacent shelves, where the box may have been placed temporarily and not replaced. If it had been a file, he would have checked the re-file area. He scanned the boxes in the bay to check if they were in their recorded or 'correct' location. Had they not been, he would have checked their recorded location, to see if the boxes there had become mixed up during the re-filing process. He looked up the transactions that were reported around the same time-stamp as the 'missing' box to see if it had become mixed up with other transactions. Failing that, he looked up the report of exceptional movements to see if there was a match for the missing barcode. In the event of any search being unsuccessful, he is required to log a 'NF-Not Found' status to the items history, for the attention of a specialised search team. It is in Rob's interest to try and find the missing item. This is a fine judgement call on his part as he cannot afford to devote too much time to finding mislaid items at the expense of retrieving other items on the pick list. The consequences of failing to retrieve an item extend to the delivery van driver, who would have to explain to the customer why their service order had not been fully satisfied.

As the pick continued, he looked forward on the list to ensure that he did not have to revisit the same location. He re-sorted the list by reference to the customer to check whether more items were required from the same box by the same customer. Furthermore, he did not accumulate the retrieved items and carry them from place to place, gathering in volume all the while. Instead, he segregated them by customer and placed them in strategic locations on each floor, near the stairwell or the cargo elevator, so that he could collect them all together and bring them to the 'Goods Inward/Outward' area of the warehouse in preparation for delivery.

In the Goods Inward/Outward area, he reconciled the quantity of items listed on the customers' Work Orders, with the quantity of items retrieved, by comparing the respective item-counts. Satisfied that all of the items required had been accounted for, he placed the items comprising each completed work order in the appropriate delivery-route bay, scanning each item again to record their latest location, i.e., in the route bay, awaiting delivery. Sensitive documents and Computer Backup Media for delivery, were left in their respective secure vaults, but segregated into the appropriate delivery route bay for the convenience of the van driver. Finally, he returned to the Warehouse Team Leader's desk to upload the latest location information from the scanner to the item history log and to receive a new assignment. Informed by the job-shadowing experience and having studied the surrounding discourse transcript, the authors refined the WPM, and numbered each different job step for reference purposes and as a precursor to a closer examination of its constituent tasks.

Detailed Task Analysis

Using another research tool viz., the Detailed Task Analysis (DTA), the authors proceeded to tease out the Picker's activities in order to disentangle and identify the MKSC that underpinned each task element and to enable each element to be linked to the appropriate level in the NFQ and each mathematics domain. This was achieved by listing each element in the sequence in which it was performed, accompanied by a description of the task as an action performed by the Picker, and separately as it may be perceived as belonging in a particular mathematics domain. The resulting Detailed Task Analysis (DTA)-Sequence, Appendix 4, shows a job-step/task number, its description and an indication of relevant mathematics domains as perceived by the authors' 'Mathematical Eyes'. That many tasks could be justified as belonging to more than one domain was immediately apparent and noted for possible implications for other cases. This reductive approach enabled the task elements of the DTA-Sequence to be regrouped by reference to their mathematics domain. The completeness of the DTA-Domain reports was verified by reconciling the quantity of tasks listed in the DTA-Sequence report, with each individual DTA-Domain report, Appendix 4, as exemplar.

This provided a sound platform upon which the MKSC underpinning a specific job could be aligned with the National Framework of Qualifications, NFQ.

Aligning the Underpinning MKSC with the NFQ

The application of the BAM, WPM and DTA tools, culminated in four, domain-specific profiles of the mathematical underpinnings of the performance of work in Case Study 1, together with a wealth of relevant contextual information. By viewing this as a portfolio of evidence of mathematical knowledge, skills and competence displayed by the worker, the researcher proceeded to apply the QQI-NFQ Alignment Protocol appropriate to each mathematics domain (full details available in Keogh, 2013). Each level of topic-complicatedness of the NFQ is realised in each syllabus as a series of Specific Learning Outcomes (SLO). An examination of each element of the DTA was linked to the relevant SLO as evidence of the relevant SLO. However, MKSC in work is deployed in the performance of work without regard for whether they constitute complete sets of SLOs as would justify the achievement of an NFQ level. This highlights another challenge for RPL. For example, an understanding of the nature of an integer is necessary at a very modest level of the recognition of Quantity and Number, whereas it may never arise in work.

The NFQ in Ireland describes the expected characteristics of the topic related learning appropriate to each level. In this way, the depth of the learning increases by increments of complicatedness in the sense of difficulty to understand and reflected by ever increasing intellectual demand expressed in terms of Significant Learning Outcomes, SLO. While this is possible within the well defined parameters of curriculum and syllabus and managed in the formal learning environment, it does not sit well with the workplaces observed in the case studies. In contrast with the formal learning environment, workplaces are designed to achieve objectives that suit the company's needs for efficiency and economy, with minimal regard for the provisions of NFQ, beyond its role in recruitment. Instead of featuring escalating levels of complicatedness, workplaces are unique combinations of factors that construe complex and sophisticated environments in which a broad spectrum of knowledge, skills and competence is used in varying degrees of intensity. These factors reside alongside the provisions of the NFQ, and are recognised as relevant or 'on the job' experience. The authors developed the Workplace Context-complexity Protocol, Appendix 5, to capture these workplace affecters and to add depth to the understanding of how formal MKSC are deployed in work.

In line with the FETAC/QQI Policy on the Recognition of Prior Learning (Further Education Training and Awards Council, 2009), the authors compared the DTA-Domain profile with FETAC(QQI)-NFQ Alignment Protocol, referencing each mathematics domain in turn, to determine whether they had been demonstrated or could be inferred, taking surrounding contextual and 'Mathematical Eyes' factors into account. In this way, the DTA-Domain profiles were annotated with the ID and level code of the Specific Learning Outcomes (SLO) so that on aggregate, the level of mathematics activity could be determined for each mathematics domain of Case Study 1. In the case of each domain for Case Study 1, it is readily perceived that while some SLOs are clearly visible or inferred, there is no evidence available for others. This is not to indicate that the worker is not capable of, or has not mastered the requirements of these SLOs. It simply underlines that mathematical activity in the workplace is acquired and used for its own internal purposes, and not with a view to claiming recognition or accreditation. In contrast, the FETAC/QQI SLOs are grouped by complicatedness, on the assumption of graduated degrees of difficulty, each building on its preceding level having been accomplished in full, in preparation for study at the next, higher level. The implied *impasse* is that the although there may be evidence of mastery in the workplace of a proportion of SLOs calibrated at different FETAC/QQI levels, full accreditation is not awarded unless all the SLOs, specified for a particular award are met – although it is possible to gather accreditation incrementally. Since the knowledge skills and competence requirements developed in response to workplace demands are not purposely aligned with FETAC/QQI learning outcomes, it may occur that some, but not all, SLOs are met at several levels, but none completely. For example the knowledge and skills to differentiate between an integer and a real number is a key SLO at fairly low levels of mathematics accreditation, but is seldom required in the workplace, especially in low-skilled, low-paid jobs. Under this constraint, only Quantity and Number, Space and Shape, and Pattern and Relation warrant accreditation at Level 1 for Case Study 1. Data handling and Chance gives evidence for only half of the SLOs at this level, even though an SLO at a higher level is clearly demonstrated (Figure 5.1).

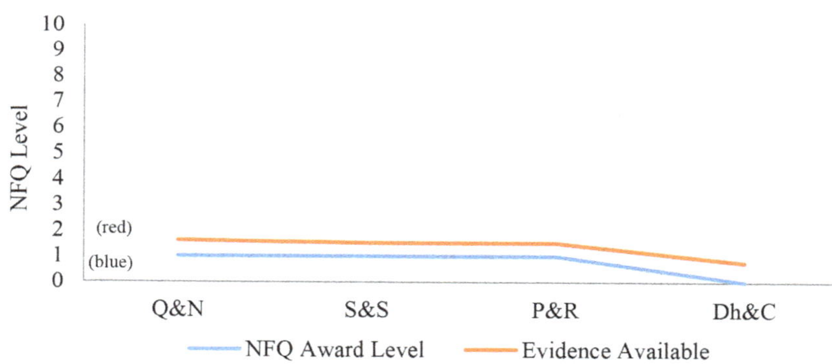

Figure 5.1. Company A, case study 1, mathematics knowledge and skills aligned with the national framework of qualifications

The Mathematical Knowledge and Skills evidenced in the Workplace DTAs may be formally recognised at level 1 (blue), in every domain except Data handling and Chance. Were it possible to recognise the mastery of only some of the SLOs at a higher level, the alignment with the NFQ would still fall short of level 2 (red), in this case.

The Competence strand of the NFQ is assumed to be embedded in the curriculum and accomplished in line with the FETAC/QQI SLOs level, although it is neither explicitly taught nor assessed. In practice, the recognition of Knowledge and Skills assessed at any given level, accredits Competence in Context, Role, Learning to Learn and Insight at the same level, even if it is demonstrably different. On this basis, and by aggregating the 4 domains into mathematics in the interest of clarity, the formal recognition of the MKSC at the service of Case Study 1, is depicted in Figure 5.2, at its most optimistic.

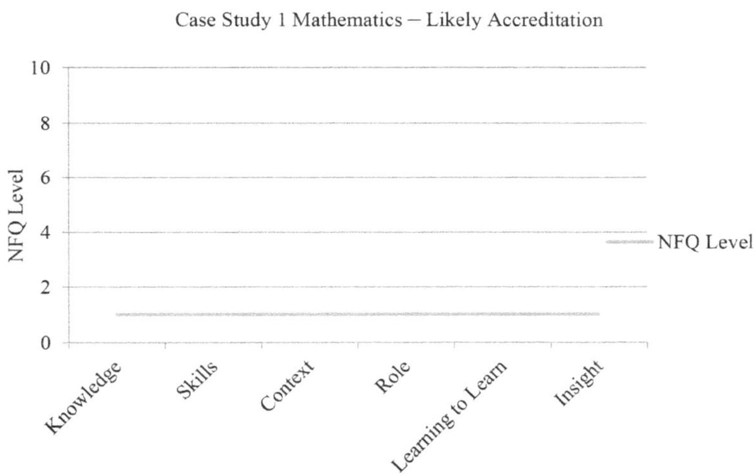

Figure 5.2. Company A, case study 1, mathematics knowledge, skills and the competence strands of context, role, learning to learn and insight – standard interpretation

Learning to Learn is described by the NFQ as occurring wholly and solely within a structured learning environment, and can be credited to Rob, at the lowest level. However, it is clear from the discourse transcript for Case Study 1, that competence, in terms of Context, Role, Learning to Learn and Insight, may not be assumed to align with the level of difficulty of the mathematics employed. In consultation with the NFQ Grid of Level Indicators 1–5 (QQI, 2012) and informed by the evidence available in the discourse transcript, the authors interpreted that the competence displayed by Rob in Context, Role , were consistent with level 3, whereas Insight was at level 6. This interpretation of the NFQ's provisions, presents a rather different, and perhaps more complete, worker's profile (Figure 5.3).

The knowledge and skills levels represented in the graph (Figure 5.3), equate with the level of complicatedness of the mathematics underpinning its constituent tasks. However, it can be seen from the Job-Shadowing of this case study that competence in context and role align with a higher order provided by the NFQ guidelines. Learning to Learn is conditioned by having to occur in formal settings, which equate

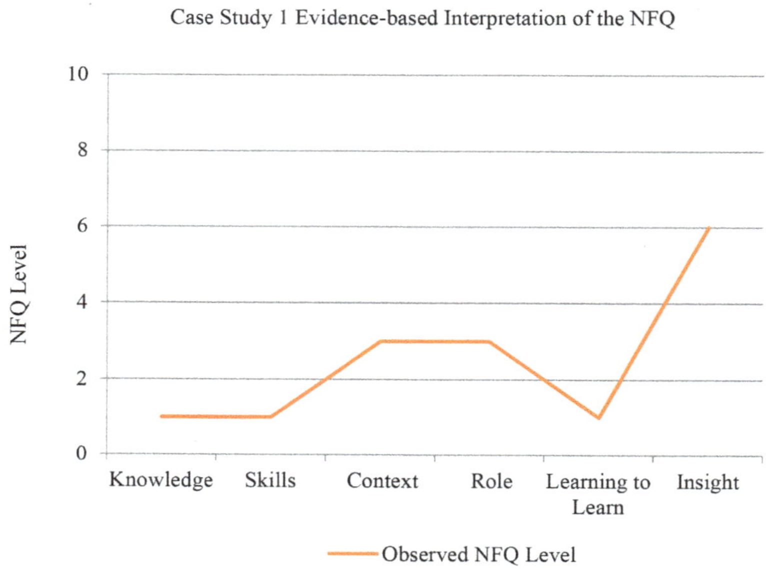

Figure 5.3. Company A case study 1, evidence-based interpretation of the NFQ provisions to represent the underpinning MKSC

with the Knowledge and Skills likely accreditation level. Striking is the degree of insight displayed by the worker, which may point to another type of competence in the workplace.

Mathematics Invisibility: 'Rob'

In the early stages of this research, the authors struggled to find an approach to deal with Mathematics Invisibility, and whether or how it might be experienced by the worker. There were several issues. The researcher was reluctant to introduce the term 'mathematics' or pointedly to refer to mathematical interpretations of the worker's actions so as not to influence the worker's discourse. Mathematics invisibility is a subjective experience, unique to the individual. It is a comparative of what is seen relative to what is available to be seen, which itself is a subjective judgement and not amenable to any absolute measure. To equate the absence of mathematical language with the absence of mathematical activity may lead to erroneous conclusions since mathematical activity may occur as a heuristic, expressed in informal, locally specific language. For example, in the warehouse, a 'tetris job' meant to re-pack a variety of shapes of goods, tightly, into the minimum amount of space, taking account of weight, stability and fragility while retaining ready accessibility. 'Rob' was able to evaluate accurately the space that would become available after a 'tetris job' had been completed, even though the area seemed to the researcher to be almost full to capacity.

Reflecting on the case study selection criteria, that the level of MKSC underpinning Case Study 1 barely registers on the NFQ is not surprising, yet that does not seem to do justice to the MKSC required to do the job. Rudimentary knowledge of the alphabet and its sequence, in addition to being able to recognise and count 2-digit numbers was sufficient to locate a single storage bay from a given starting point. Doing this continuously, connecting multiple nodes or collection points, in multiple dimensions, given a number of constraints, under pressure of time and precision, involves something rather more than the application of simple skills and uncomplicated knowledge in a simple context. Resolving 'missing item' problems, required him to place himself in another person's position to imagine how an error might have occurred, in a highly controlled system. The requirement to be precise carries with it the possibility of both immediate and delayed consequences. Misplacing an item amongst several millions of similar items may make it impossible to find, with possibly serious consequences for the customer and the employer. Replacing property deeds, that had been lost by the company for example, could cost half a picker's annual salary, the cost to the company's reputation could be many multiples more. That the picker is ranked amongst the lowest status in the company does not relieve him of the responsibility of having handled the item last. The point is, that while the MKSC that underpin the work practice may be aligned with the NFQ a relatively low level, it seems that the context may have a quite significant impact on the way in which they are used. Similarly, many jobs may use relatively low levels of MKSC expressed in contrasting contexts. It may be that calibrating the underlying MKSC in terms of degree of difficulty alone fails to capture other, crucial, factors which may say something about what the worker knows, as distinct from what s/he does, and that that should and could be recognised formally.

Providing the driver with an incorrect or incomplete inventory for delivery, would necessarily delay the driver, and perhaps place him/her in a difficult position in relation to the customer. Providing the incorrect computer backup media to the customer, at the very least would inconvenience the customer, and cause the driver to abandon his delivery schedule, to return to base to resolve the problem and re-deliver the correct item, thus extending the driver's day unnecessarily. That the driver is required to complete his day'swork, however long it takes, may result in interpersonal conflict in the workplace and all that that entails, and may persist over an extended period.

Case Study 2 – Van Driver 'Jim'

Employed by Company A, Jim described himself as 'only a van driver'. In consultation with the Transport Manager, he was selected as a suitable case study because his main function comprised the delivery of items that had been retrieved from storage in the warehouse to the customers who requested them. He may also be required to collect items for storage in the warehouse from customers on request. His delivery route is busy and mainly concentrated in central and south-central Dublin

City. He receives his instructions in the form of printed delivery/collection Work Orders, extracted from the Work Order Log, by reference to customers' assigned route code. He is equipped with a handheld scanner/data device, to which his workload can be downloaded as in the case of Rob, Case Study 1. Additionally, work can be assigned to him by the Transport Manager to relieve the pressure on other delivery routes. In order to meet the morning delivery service requirement, Jim's working day begins at 7.00 a.m.

Contextual Factors

There are dozens of customers located in Jim's delivery route, each requiring different frequencies and types of service. They may fully or partially occupy one or several buildings. Many offices are located in Georgian Buildings, which, as a legacy of the early 19th century architecture, feature narrow staircases especially to the upper floors, and seldom provide an elevator of any kind. They differ in their access and security protocols, provision of delivery bay, parking, and formalities, which on occasion conflict with Company A's Standard Operating Procedures (SOP), custom and practice. They may be situated where legal parking is conditioned by time of day and/or side of the road. In the case of computer backup media, delivery and collection may need to concur for the sake of efficiency, but may be subject to the most recent backup being ready for collection. Traffic density in the city fluctuates to reflect school terms, industry specific fixed holiday periods, road works and the prevailing weather conditions – rain tending to increase traffic volumes and reduce pedestrian awareness. Customers are continuously canvassed by rival document management companies offering competitive pricing and service guarantees. This places Jim in the 'front line' to maintain customer satisfaction and to cultivate a measure of loyalty by providing an efficient, professional and friendly service. On occasion, Jim may be asked to collect items for storage for which there is no Work Order or that is not addressed in an existing Service Level Agreement (SLA). While Jim may exercise some latitude for long-standing customers, providing a service outside an SLA could create difficulties for himself and his employer. At the same time, declining such requests can annoy the customer and give rise to complaints which are subject to disciplinary proceedings. Jim's van is fitted with a GPS transponder, which reports his exact position, speed, engine status, and dwell time at any location. There are a number of designated 'red' zones, including around his home, which he is not permitted to enter without prior permission. Breaches of this condition are automatically reported to the Transport Manager, who may initiate disciplinary action at his discretion.

Job Shadowing – Case Study 2, 'Jim'

The Work Practice Model, for Company A, Case Study 2 was prepared with the benefit of having job-shadowed Jim for several hours in the course of his work, and having studied the transcript of the surrounding discourse.

Jim's workload comprised collections/deliveries for customers located in the part of the city assigned to his route. He received his instructions in the form of printed, itemised Work Orders that had been extracted from the Work Order Log and assigned to him by the Transport Manager. In addition, he downloaded the item details, including item barcodes and their locations e.g., route bay or vault, to his handheld scanner/data recording device. The Work Orders determine Jim's work objectives for each day and shape his work practice. They comprise itemised deliveries and collections, scheduled for completion either before or after noon, scattered throughout the geography of his delivery route. While there are regular and frequent customers, the combination is never the same, meaning there can be no fixed, optimal route. Included in the day's work was an early delivery for a customer whom Jim knew would be serviced by the 'bulk truck' later the same day. The 'bulk truck' service collects and delivers quantities of storage boxes that exceed the capacity of the regular van. In an effort to minimise his workload, he contacted the 'bulk truck' crew to negotiate the handing-off of his Worker Order to them. It transpired that they were either unwilling or unable to oblige him. He used his knowledge of the customers to determine flexibility in the delivery/collection time, before segregating the Work Orders into two separate piles i.e., a.m. and p.m. He then concentrated on the a.m. pile, sporadically re-ordering them into viable delivery sequences, evaluating the implications, taking a range of variables into account. He explained that there were no official guidelines as to how to work out the most efficient route and that every driver had their own way. He preferred to arrange his customer calls in a loop, finishing back at the base in time to unload the van of collections, have lunch and to prepare for the p.m. schedule. This meant that he would pass customers' premises on the outbound leg of his route, because they were located on the side of the road that suited the return leg. He factored in that some computer backup media might not be ready for collection in time to suit his ideal schedule, and that several locations in the city permitted vehicular traffic or parking within strict time limits. Finally, having taken into consideration one-way streets, *cul-de-sac*, traffic congestion due to road works, traffic volume due to schools' term time, and the weather, he settled on an initial delivery schedule which was capable of responding to change.

The delivery schedule determines the loading sequence of the van on a 'Last In – First Out' (LIFO), basis in principle, but this must be tempered by the stability of the load at the outset and take account of the, as yet, unknown return load. Stability is crucial, since, should boxes of documents topple over while in transit, they may become mixed together and require extensive efforts to re-sort them, at least delaying the delivery and disappointing the customer. The 'rule of thumb' is that deliveries are stacked at the rear doors of the van, collections at the side door, but this adds to instability in the event of a sudden stop.

Taking each work order in turn, Jim checked that all of the specified boxes, files and documents were located in his delivery route bay. It seemed that two boxes were missing. He consulted his handheld scanner to check that the boxes had been retrieved from the warehouse and placed in the delivery route bay the previous day, in case there had been some oversight or amendment. Thus assured, he searched adjacent bays, and found the missing items, explaining that on occasion, boxes are moved for any number of reasons and not always properly replaced. He did not move them back to his own route bay, instead making a mental note of their location. Subsequent events would prove the folly of this decision. Since the cycle of computer backup media features on a daily basis, he made his way to the media vault, bringing a small trolley for convenience, and being conscious of his occasional back-pain. The Picker had previously segregated the tapes, drums and disks into an area of the vault set aside for delivery bays. Jim selected computer media transport case, and scanned the item 'into the case' to record their new location as 'in transit'. Different sized and shaped media were stored in appropriate transport cases, and two items were already contained and locked into a specially designed container. The media transport cases were loaded onto the small trolley and brought to the delivery-route bay, having collected some deeds from the document vault along the way. Jim assembled two document storage cartons, from their flat-pack form, in order to contain loose files and documents for security in transit. While the company provides large plastic containers for this purpose, Jim refuses to use them because he considers them to be in breach of the health and safety guidelines for the safe handling of heavy and relatively wide containers. This view brings him into conflict with the Transport Manager; Jim's primary concern being that should he be unable to work as a result of handling these plastic containers, the company would not pay him or his medical bills.

Now that all the items for delivery had been verified as having been gathered in the route bay, Jim reversed his van into the Goods Inward/Outward area and loaded the deliveries ensuring stability and LIFO. He selected the status 'truck validated' from a menu of options on his handheld and scanned each item from the delivery bay 'into the van', in preparation for effecting the deliveries/collections specified on the customer's worker orders. Whether distracted by the presence of the researcher, it later transpired that he did not load the two boxes that had become misplaced in an adjacent route bay. Before leaving the warehouse, he checked with customer service to see if there had been late changes. While it is the responsibility of customer service to inform Jim, he takes the initiative because he bears the added inconvenience of having to retrace his steps if late changes are communicated to him while he is *en route*. He also collected a wireless printer so that he could issue receipts to customers.

The morning rush hour had already begun by the time Jim started out from the warehouse depot. He passed several customers on the basis that they were either on the 'wrong' side of the road or would not be ready. On arrival at each customer, he used the facilities provided or the most convenient, legal parking place, taking note of the time conditions particular to loading activities, even though these were often flouted by common practice of other road users. He selected the relevant Work Order and loaded a hand truck with the boxes for delivery, scanning each item to record a 'delivered' status. For the delivery of computer media, he brought the entire media transport case to the customer and scanned them as 'delivered' when they were handed over to the customer. These procedures deviate from Standard Operating Procedures (SOPs), having been modified by Jim as a result of his worked experience. Since the SOPs were created and maintained by individuals who have no experience of doing the job nor of the variety of customers requiring special treatment, Jim feels obliged to use his 'common sense', enhanced by experience to provide the customer with the service requested while minimising the risk of error. When the hand-over was complete, the customer signed the printed receipt and was provided with a copy, detailing the barcodes of the individual items delivered. The signed receipt was attached to the original Work Order, as evidence of completion. Items being returned to storage were scanned as being 'collected' from the customer, receipted and loaded in the van. In the case of items that were new to Company A, Jim issued new barcodes and attached them to items, recorded them on the scanner and printed a receipt for the customer, before returning them to the van. In one case, there were many more files to be collected than originally ordered by the company. Jim made a judgement about whether the van would have sufficient capacity to accommodate the extra items, along with the volume of items scheduled to be collected as compared with the risk to the relationship of refusing as instructed by the SOPs. Each customer visited had its own rules of access, and sensitivities to which Jim made an appropriate response.

The experience of a particular customer alerted Jim to the likelihood that the computer media due to be collected might not be ready. Because this had been a frequent occurrence, the customer-contact person had undertaken to inform Jim by phone when the item was ready for collection. This was a welcome, but unofficial arrangement not sanctioned by the SOPs, but it did save time. Due to recent changes in personnel, the promptness of making media available for collection at the agreed time had deteriorated, passing the initiative to Jim to make contact by phone to check on the customer's readiness for collection, when the pattern of his delivery route suited. However, it was incumbent on Jim to attend the customer in person as required by the Work Order, whether the media was ready for collection or not, because it was a billable transaction. During his shift, Jim was assigned extra collection duties by the Transport Manager, causing him to review his planned schedule to accommodate additional locations. On arrival at one customer location, it became apparent as a result of comparing the Work Order with the items for delivery, that some of the boxes ordered by the customer had not been loaded onto the van. He checked the contents of the van without success. His handheld scanner reported that the 'missing' items had been 'truck validated', i.e. loaded onto the van. It was only then he recalled the items that had been misplaced in an adjacent delivery route bay and that he had neglected to return them to their proper location before loading the van. He first contacted the warehouse to verify that the goods were still there. He also remembered that he had a call scheduled for the customer in the afternoon shift. Rather than arrive with an incomplete delivery, drawing attention to his error, he decided to include all of the boxes in the afternoon delivery.

The job-shadowing had caused a delay of about 30 minutes to the driver's schedule, who was already under pressure due to a heavy workload. On completion of the national survey having fed back to the driver the breadth of his Numerate Behaviour, the driver, sitting taller in his seat, handed his mobile phone to the researcher and said, "would you ring my wife and tell her I am not stupid".

On his return to base, Jim unloaded the van and scanned each item into the re-file area, in readiness for the warehouse team to relocate files in their original storage boxes, secure deeds and computer media to their respective vaults, and boxes to whatever space was available. He uploaded the contents of the data scanner to update the items' storage and movement history log and returned the annotated Work Orders, together with their attached receipts to the Transport Manager. The richness of the discourse which accompanied the job-shadowing of the van driver, permitted the drafting of a more precise WPM, Appendix 3 as exemplar.

Detailed Task Analysis

In keeping with the process applied to the Picker, the researcher produced a Detailed Task Analysis, Appendix 4, identifying the underpinning Numerate Behaviours in the sequence in which they occurred, indicating their appropriate Mathematics Domains, as perceived through Mathematical Eyes. This enabled each task and subtask to be re-grouped according to Mathematics Domain, resulting in DTA for each of Quantity and Number, Space and Shape, Pattern and Relationship, and Data handling and Chance in preparation for the alignment stage.

Aligning Underpinning MKSC with the NFQ

The same QQI-NFQ Alignment Protocol (full details available in Keogh, 2013), as in the case of the Picker, was applied to the DTA-Domains for the Van Driver. The formal recognition of prior learning, in keeping with the existing policy, permits the accreditation of Jim's use of MKSC in work at level 1 (blue line), even though higher levels of Knowledge and Skills were demonstrable (red line) (Figure 5.4).

As with Case Study 1, and in keeping with the practice of assuming that the level of NFQ Competence is determined by the recognised levels of Knowledge and Skills, the workplace MKSC profile for Jim is set at the level at which the appropriate SLOs are met, rather than reflecting many aspects of the variety of context in which his job is set (Figure 5.5).

In contrast, an interpretation of the accompanying discourse, taking the range and variety of contextual variables into account, is more fully represented in Figure 5.6, based on evidence provided.

Whether the difficulty of the mathematics knowledge and skills are more accurately assessed at level 1 or 2, is not the central point here. Rather, it is that the contexts in which they are deployed are multiple, variable and unpredictable. Jim occupies a role that has responsibilities to his employer, co-workers, and fellow road

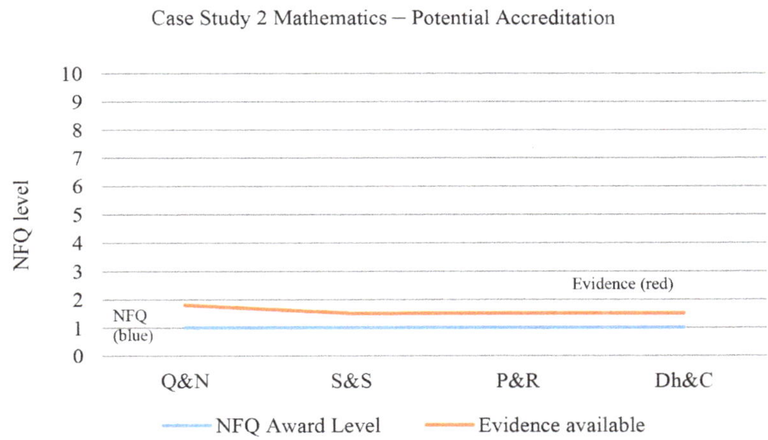

Figure 5.4. Company A, case study 2, mathematics knowledge and skills aligned with the national framework of qualifications

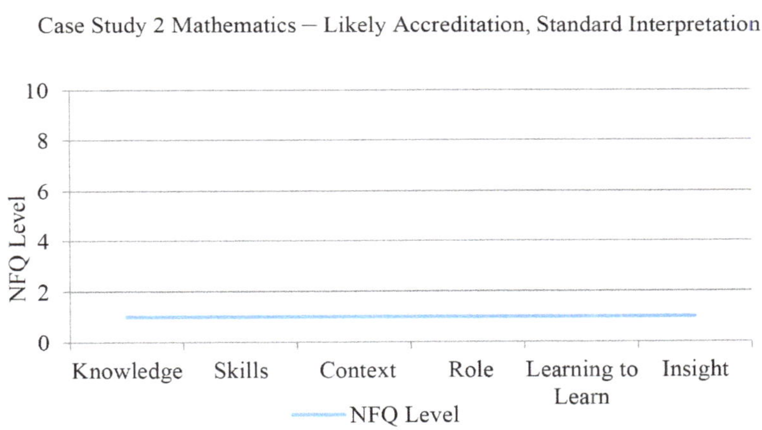

Figure 5.5. Company A, case study 2, mathematics knowledge, skills & competence

users. He is required to adopt different approaches to suit different customers, their requirements and regulations, even when they are in conflict with his employers standing instructions, in order to maintain a competitive edge in the face of stiff and persistent competition. His insights are highly developed in contact with customers and in reconfiguring his delivery pattern in response to volatile conditions. Even though the formal conditions imposed on Learning to Learn restricts this aspect of Jim's Competence profile to level 1, interpreting these contextual factors within the provisions for Competence in the NFQ Guidelines (QQI, 2012), suggests a more complete profile of the MKSC displayed in Case Study 2 (Figure 5.5).

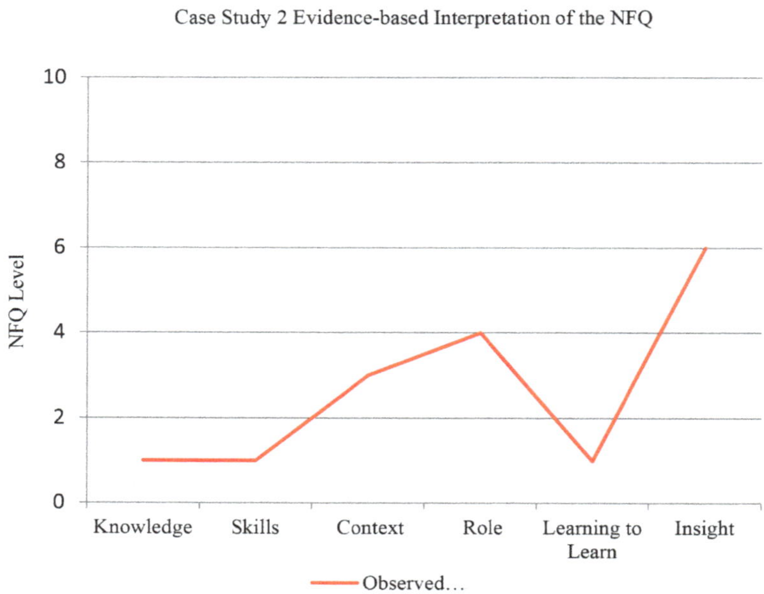

Figure 5.6. Company A case study 2, evidence-based interpretation of the NFQ provisions to represent the underpinning MKSC

Mathematics Invisibility: 'Jim'

There was very little overt evidence of the use of mathematics in the course of this case study. While Jim identified, evaluated, acted upon and communicated about mathematics containing information, in order to manage his work, and solve problems, these facets were not easily distinguished from other information contained in the performance of his work. The use of mathematics terms was present in the discourse, but in a sense loaded with social rather than mathematical meaning – or so it seemed. The extent of his acknowledgement of the mathematics dimension to his work was not palpable, even though it was visible to 'Mathematical Eyes'. His attention was on meeting his objectives, in the optimal amount of time, albeit supported by multiple, highly practised routines and thought processes, honed over the years to the point of being automatic operations. Similar to Case Study 1, the role of MKSC in Case Study 2 was accurately pitched at the lowest levels of mathematics when described in terms of difficulty alone. However, there was corroborating evidence that low level mathematics knowledge and skills alone, does not equip the worker to do their job, because of the range of factors that influence how that MKSC is expressed. This was an early herald of the absence of a framework with which to calibrate and talk about the multiple, unpredictable facets that impinge on the deployment of MKSC in the workplace A, and perhaps others too.

RESEARCH HOST COMPANY B

Commercial Purpose

Company B supplies electronic components to the communications industry in the wholesale, rather than the retail, sector. Its services create and programme devices to meet specific requirements and/or re-programme generically manufactured computer switches and plug-ins to customers' specifications. It imports more than 1200 different device types in bulk and exports them in smaller quantities on demand. The accuracy of the demand-forecast is crucial, and must balance the extent of the supply lead time, with the risk of having out-of-date devices in stock for which there is limited and diminishing demand. The value of seemingly identical components ranges from a few cents each to more than €1,000, depending on their functionality and internal programming. The overseas operations are treated as internal customers and may be required to re-supply sibling organisations under the conditions of being over-stocked or more advantageously positioned logistically. Their reputation for precision is core. They must 'get it right' first time and every time since the cost of error can be hugely disproportionate to the type of error. For example, the supply of an incorrect component to a customer, even though it may be functionally identical, may alert them to a cheaper alternative, which may have the effect of damaging customer relations and profit margin. This is a dynamic industry where standards change rapidly and customer loyalty is liable to seek lower cost economies. Profit generated by the quick turnover of purchases to sales can be undermined by slow paying debtors, which in turn restricts cash flow and inhibits supply, which can be catastrophic for this type of company.

Business Activity Model

The Business Activity Model (BAM) for Company B was refined, finalised in discussion with the Finance Manager, Customer Service and Warehouse personnel, producing a model that reflected their different yet coherent perspectives; full details available in the original thesis.

As might be expected of a small operation, albeit with global reach, the business activities depicted are very well integrated and interdependent in a matrix rather than silo structure, especially regarding Purchasing and Stock, each of which are closely monitored by Finance Management. Purchasing decisions are the responsibility of Finance Management, supported by Customer Service and influenced by the sales volumes reported by the sibling companies and the consumption trends that they indicate. However, Purchases can only be completed so long as Company B is in good credit-standing with its suppliers. This requires payment for previous shipments within the agreed credit terms. The capacity to meet payments schedules is directly influenced by the company's ability to meet the orders to supply their own customers, both internal and external, and to collect the monies due from their debtors in a timely fashion. Every precaution is taken so that Company B's customers have no grounds for delaying payment of their invoices e.g., through error or omission. Orders from customers who are in default of their credit terms as determined by Finance Management, are delayed or not processed, pending settlement of their account. Company B maintains its liquidity by assiduously attending to collecting monies due from their customers before having to settle accounts with their suppliers. Otherwise, the burden of financing the capital overdue from debtors pares back the potential for profit. In sum, Company B must predict the expected orders from customers, secure the relevant supplies on credit, fill the orders accurately and deliver them on time, collect the money due and pay suppliers before the next cycle can begin. Key in this setting is the interaction between the Customer Service Personnel and Warehouse Operatives. That the latter appeared to be dominated by manual handling, and the former featuring interpersonal skills, seemed to offer two ideal case studies for the apparent absence of overt mathematical activity.

Case Study 3 – Customer Service – 'Laura'

Laura, having agreed to participate in the research, described her job as 'Customer Service', which, on first contact seemed to encapsulate her function. She shared the telephone answering duties with an adjacent colleague, redirecting callers to individuals within the company. As the researcher sought to clarify the BAM and to develop a WPM for the role of customer service, the extent of Laura's knowledge of, and involvement in, the company and its activities began to emerge.

Contextual Factors

Company B has developed manufacturing and supply facilities in 13 countries in order to compete in a worldwide competitive market. While the overseas offices are largely self-sufficient, they collaborate to optimise their purchasing power in negotiating price and to minimise their exposure to slow-moving stock, with the Dublin office at its hub. Company B has negotiated exclusive agreements with a relatively few manufacturers, to supply particular, although not unique, components to the market. Potential buyers are scattered throughout Europe and the rest of the World. The customers that Laura deals with range from global companies that are household names, to substantial manufacturers and consumers of electronic communications hardware in virtually every country in the World. Given the distribution of time-zones involved, Laura's job is not confined to regular office hours.

Consequently, time is of the essence, supported by her command of the current stock positions in terms of type, quantity and cost price, together with the relative cost of effecting delivery of the same components to the same customer from different parts of the world. She has worked with Company B from its modest beginnings and, with secretarial training alone, has developed and configured a standard accounting software package to meet their stock control and order processing needs. While she has no responsibility for collecting the company's income, she is pivotal in enabling transactions to be completed so that revenue can be maximised at key points in the company's financial cycles.

Job Shadowing – Case Study 3, 'Laura'

Job shadowing Laura was particularly challenging due to constant interruptions while she answered the telephone, fielded queries from customers, suppliers, courier services and from work colleagues. It is a very dynamic role, both proactive and reactive, in which different parts of different functions are interleaved in unpredictable ways, perhaps several being concurrent. This is a major dimension to her job, she may not, and did not suspend all other activity while she completed a single task from 'end to end'. From amongst this range of discontinuous activities, the author pieced together three related elements that constituted a substantial part of her job viz., Purchasing, Goods Inwards and Order Processing as depicted in her Work Practice Model (WPM).

Regarding the acquisition of stock from suppliers, Laura extracts stock reports from a computer supported system which she designed and maintains to reflect the current stock position. A piece of stock is identified by a 15 digit alphanumeric code which indicates its origin of manufacture, function, lead content, wavelength, frequency band, optimal operating-temperature, signal capacity, and other functional information. It also bears the manufacturer's part number. These identifiers are not unique to the individual item, but to a device type. Stock records maintain quantities of devices, quoting their identifier codes and segregating them as to their status code number which indicates whether on 'back order', 'free', i.e., available for sale in-house or in another location, 'committed' to an incomplete order, 'on-hold' awaiting payment, taken as samples, on 'test', or 'in quarantine', i.e. in dispute. She also reviews each part to establish a 'run rate' which estimates the level of demand and dwell time in stock. This is synthesised with regular reports from their main suppliers regarding expected/imminent shipments. Then, in consultation with Finance Management and Marketing/Sales, the future stock requirements are forecast, including a buffer stock, enabling Laura to issue a Purchase Order, log it on the computer's accounting system, and to update the Stock records. While Laura states that 'the computer does everything', she adds the caveat that 'we have to check everything as well', which suggests that the data retrieved from the computer system may be nuanced in some way that cannot be recorded and that is known only to a few. At the end of every month, Laura prints a 'stock on hand' report and orders a 'Stock Check' to be carried out by a Warehouse Operative to verify its accuracy by actually counting each listed item.

Goods Inward describes the processes stimulated by the receipt of deliveries from suppliers, having been previously ordered. The courier services, for their own convenience, tend to deliver small packages directly to Laura and large packages directly to the Warehouse. Every inward shipment is accompanied by the supplier's 'packing slip' which is checked by the Warehouse Operative (Case Study 4), annotated as to errors and omissions and returned to Laura. Having resolved the discrepancies, which are rare, Laura updates the stock records, changing the status from 'On order' to 'Received into stock', thereby validating the original Purchase Order, upgrading its status to 'Bill Payable' for the attention of Finance Management.

Order Processing deals with orders to supply customers that are located all over the World. Typically received by email, customer orders detail the device type description, ID code and part number in the desired quantities. They may include additional billable service requirements e.g. re-programming devices to behave in a particular environment, and packed to their specifications, especially if they are for re-sale to other supply chains. Some customers insist on the original part number being replaced with a number of their own design, and that Company B delivers the goods directly to a third party.

Laura 'loads' a customer order on to the computer stock system. Should the stock be 'free', i.e. in the warehouse not 'free of charge', it is committed to the customer order. If not, she looks up the stock to see if the goods are available in one of the company's other locations, whereupon, she will optimise the delivery to the customer taking account of the relevant logistics and associated costs. She will always seek to fill a customer's order from the company's local or extended resources, before issuing a purchase order or including it in the demand forecast if the lead-time permits.

Having ensured that the customer is in credit, and that the goods ordered are 'free' and in local stock, Laura issues a Delivery Slip, replicating the customer's specifications to the Warehouse Operative, together with an expected completion time. She logs on to a courier service website to book the shipment for collection, providing details of the goods, delivery address, insurance, and customs formalities consistent with the transit and destination jurisdictions. The courier company issues the appropriate shipping documentation, which Laura passes to the Warehouse Operative. In the event that the goods are more economically supplied and delivered by a sibling operation in another location, Laura directs the customer's order accordingly. Once the goods have been collected by the courier, they are deemed to have been dispatched, the stock records are updated, automatically noting that the order had been completed and released to the accounting operations for customer billing.

Detailed Task Analysis

The combination of Job-shadowing activities, the resulting WPM and the surrounding discourse enabled the author to pick apart Laura's work practice to reveal the underpinning MKSC. Case Study 3 differs from Cases 1 and 2, in that it was largely deskbound, and dependent on computer and telephone use. While this setting enabled multiple facets to co-exist and the rapid switching of attention between them, the researcher assembled the observed job-fragments into three separate, yet interdependent components viz., Purchasing, Goods Inward, and Order Processing. The relevant Detailed Task Analyses, in the sequence in which they occurred are available in the original thesis but follow the same template that the example in Appendix 5, segregated into their various Mathematics Domains as observed through the researcher's 'Mathematical Eyes'. A tally of the incidence of tasks is maintained at the foot of each DTA, for reconciliation purposes. As a prelude to the alignment of the underpinning MKSC with the NFQ, the DTAs were then grouped according to Mathematics Domain. Full details of the identification of detailed tasks, their underpinning mathematics concepts and their subsequent mapping on to the NFQ is provided in the original thesis.

Aligning the Underpinning MKSC with the NFQ

The application of the QQI-NFQ Alignment Protocols for Quantity and Number, Space and Shape, Pattern and Relationship, and Data handling and Chance, and in consultation with the case study discourse transcript, individual tasks MKSC components were aligned to the NFQ. In keeping with the trends emerging from preceding case studies, failure to detect evidence of individual SLOs, inhibits the recognition of the level of MKSC that may be accredited to the fullest extent. The competence associated with levels of accreditation seemed to focus on the dimension of complicatedness, to the exclusion of others. Furthermore, the requirement to describe natural numbers, as distinct from real numbers or integers, while an essential building block for further study in mathematics, is irrelevant in the workplace, yet it is a stipulation of NFQ Level 3. The contrast between the level of expertise observed, and that for which accreditation may be awarded, seemed to point towards the structure of the NFQ being more suited to a formal learning environment than that of the workplace (Figure 5.7). While permitting partial credit may be useful to indicate progress toward the next level, it is crucial that a full award attests to the achievement of a body of work that is in keeping with the national standards and curricula. The MKSC deployed in the workplace, does not to fit well the provisions and conditions of the NFQ, attracting limited accreditation potential only.

The tendency of the NFQ accreditation implementation mechanism to assume that Competence in Context, Role, Learning to Learn and Insight, occurs in equal measure to that of the Knowledge and Skills, offers a profile of the MKSC underpinning Case Study 3 in a way that seems to exclude important factors for which there is ample information (Figure 5.8).

In contrast, interpreting the NFQ Competence guidelines and applying them to Case Study 3, independently of the levels of Mathematics Knowledge and Skills, offers a more informative profile (Figure 5.9). That Competence in Context, Role, and Insight may be observed in the performance of work at levels that exceed that of the underpinning knowledge and skills is clear from the workplace observations. The guidelines relating to Learning to Learn, seem to constrict the recognition of this dimension to that which occurs in a formal learning environment only. There

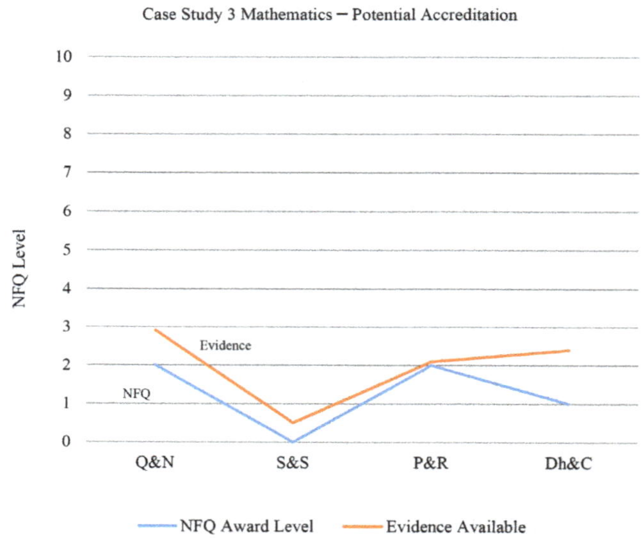

Key: Q&N = Quantity and Number S&S = Space and Shape
P&R = Pattern and Relationship Q&N = Quantity and Number

Figure 5.7. Company B, case study 3, mathematics knowledge and skills aligned with the national framework of qualifications

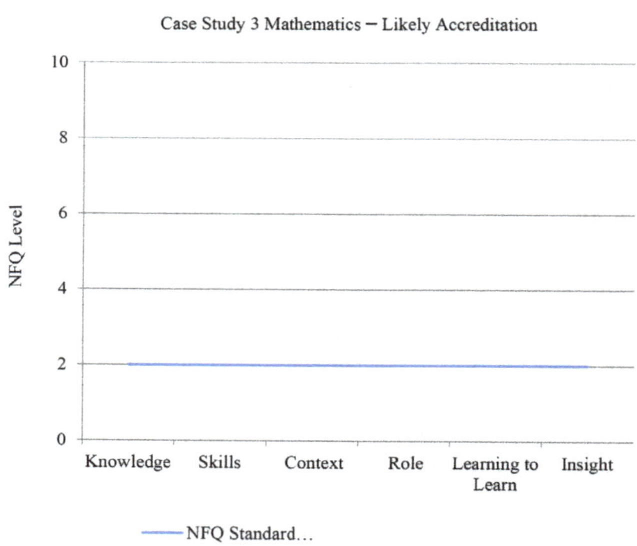

Figure 5.8. Company B, case study 3, mathematics knowledge, skills and the competence strands of context, role, learning to learn and insight, – standard interpretation

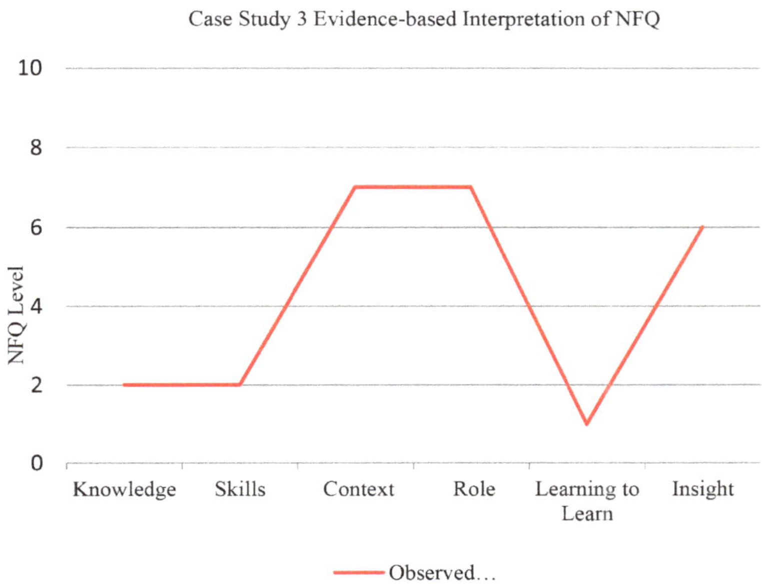

Figure 5.9. Company B case study 3, evidence-based interpretation of the NFQ provisions to represent the underpinning MKSC

is evidence that the majority of the individual's learning in this case study occurred on-the-job and out of necessity, seemingly unsupported by any formal provision. It is clear that the recognition of learning by tacit rationality and analogical thinking is beyond the provisions of the NFQ. Furthermore, the custom of aligning the level of Competence with that of the Knowledge and Skills does not appear to be operable in the context of the workplace.

The outcomes reported thus far, while emerging from three very different work practices, converge nonetheless. There is ample evidence of mathematics knowledge, skills and competence underpinning work, albeit at relatively low levels when aligned with the NFQ. It is clear that the mathematics deployed is in the service of work and may imply higher levels of achievement in mathematics without having completed lower levels as would be the norm in a formal learning environment. In addition, the combination of MKSC are more fully characterised as having a spiky profile, rather than a flat, single level of achievement in all dimensions.

Mathematics Invisibility: Laura

The processes and procedures supporting Laura in her work evolved to meet the expanding needs of Company B. The stock recording system was developed by her, incrementally, without any relevant formal training, and as a result of her curiosity to see what the software could do. None of the procedures are documented. They are multiple and interleaved, she is "sure there is a lot more to it but she just can't think at the moment". Her role is pivotal in maintaining the rapid flow of transactions and money. She attends to duties by calling to mind whatever cognitive resources as are required, including mathematics, each one an indivisible facet of the same thing – her job. This set of circumstances, especially in the light of preceding case studies, suggested to the authors that herein may lie a possible explanation for Mathematics Invisibility.

Case Study 4 – Warehouse Operative – 'Mike'

The selection of a Warehouse Operative to participate in this present research was subject to the same criteria as previously, i.e., the superficial absence of mathematics activity. In addition, profiling a second person engaged in warehouse work, offered the opportunity to compare two similarly described jobs. It is a physically demanding job, involving lifting and carrying loads of up to 20 kilos when taking delivery of cartons of goods, and preparing packages for dispatch. Closer examination revealed activities that were rather more complex and carried out under pressure of time and accuracy, involving multiple combinations of a range of components. The Warehouse Operative in this case is required to unpack and verify the receipt of electronic components and communications devices into stock, to fill customers' orders in accordance with their detailed specifications and make them ready for dispatch, and to reconcile the stock 'on-hand' with that reported by the computerised stock control system.

Contextual Factors

Mike began his career in Company B in their production department. Over the last decade, many of their labour-intensive functions have been relocated to lower cost economies overseas, with the effect of reducing the warehouse personnel to two people viz., Mike in the role of Operative and 'Ro', in the role of Warehouse Manager, currently being extended into Sales.

Cartons of goods arrive in the warehouse frequently and unannounced, although they may be expected or urgently required. The contents of cartons may range from a few items to several hundred, they may be marked as being part of a larger shipment, and may display evidence of having been mishandled or damaged. The nature of the modern supply chain is that a consignment is checked at each transit stage, so that fault can be attributed and remedied more promptly. Consequently, it is incumbent on the receiver to check that goods have been delivered as described in the accompanying shipping documents, thus relieving the previous handlers of their service obligations. The contents of the carton are described by a 'packing list' provided by the supplier. Goods received and checked are placed in stock. The 'packing slip', annotated with any anomalies revealed when being unpacked, is passed to Customer Service/Goods Inwards procedures.

Orders to supply, prepared by Customer Service, detail the quantity and type of items and how they are to be programmed, labelled, packed and dispatched. There are more than a thousand different types of components and several hundred reprogramming tools to choose from. A typical order may call for the supply of dozens of the same item type, to have the original part number label removed and replaced with a label produced in the warehouse to a given specification. The same reprogramming tool may be applied to each item in the customer's order, but may have to have a unique serial number input by the worker to a specific location within its EPROM, to act as a computer media access code. There are four different types of interface sockets used in the reprogramming function depending on the physical size of the device. According to the Finance Manager, the cost of such devices can range in cost up to "a couple of grand" each, using a colloquial rather than a mathematical quantity descriptor.

The monthly Stock Check is a critical checkpoint for the computerised stock records. Customer Service produces a list of items in the alphabetical order of the item ID code, and the quantity that should be in stock. Mike checks every item, taking no 'shortcuts' in the knowledge that he is liable to be questioned in the event of a stock shortfall. There is no documented procedure for carrying out the Stock Check. Mike developed his own strategy gradually over the years. It is supported by his practice of keeping an unofficial tally of starting quantity, additions and removals so that the quantity in stock is noted on the exterior of the boxes in which they are stored. This quantity is not absolutely reliable because, in busy periods, he may not take the time to maintain this running total. Some items may be removed from stock and provided as samples or used in testing, without the displayed quantity being modified. The Stock Check procedure takes about three hours to complete during which time Goods Inwards and Order filling activity is suspended, unless exceptional circumstances arise. Mike's warehouse colleague, while not himself a candidate for job shadowing, provided extensive information regarding the company, its structures and evolution, and made a substantial contribution to capturing the context in which Mike works. Furthermore, each person can contribute to the other's work e.g. 'Mike could select the items from stock and re-programme and re-label them, while 'Ro' could finish the order by packing them. This shared responsibility is recorded in a process control check document. These factors, together with the relevant discourse transcripts, enabled the development of the Work Practice Models that encapsulate and define three aspects of 'Mike's job viz., Goods Inwards, Order Filling and Stock Check WPM.

Job Shadowing – Case Study 4, 'Mike'

Mike's job comprises three separate yet interdependent activities. The performance of Goods Inwards tasks are initiated by the delivery of a package or carton(s) of goods to the warehouse. Order Filling is initiated by a Customer Order, produced by Customer Service and passed through to the warehouse for attention. The Stock

Check is a monthly procedure initiated by Customer Service. Any of these activities may be permitted to interrupt each other as Customer Service adjusts priorities in response to a variety of circumstances. The Job Shadowing reports the sequence of steps involved in each work component, even though they may not have occurred contiguously.

Goods Inwards

Mike was engaged in filling an Order when he was made aware of the arrival of a delivery that required his attention. The courier handed over a small quantity of cartons of different sizes and weights. He checked the consignments' details and signed his confirmation on the courier's handheld device. Had there been a discrepancy, he would have noted it as a condition of his signature. Had there been any evidence of damage to a carton, he would have photographed it as evidence for a possible future insurance claim. He brought the packages into a space in the middle of the floor and grouped them by the content of their accompanying consignment notes. Taking each carton separately, he stripped its contents and grouped the items by size and type and reconciled them with the accompanying 'Packing Slip', by identifying each item code and counting the quantities in which they were present. It is worthy of note that components such as these are shipped in 'clam-shell'-style plastic packaging, moulded to the shape of each item of its contents, so that they may not move within the packaging, while in transit. Each package is arranged in 3 or 4 rows of 7, 8, or 9 item-shaped depressions, depending on the item size, the quantity shipped and the suppliers' preferences. A package need not be completely full.

Mike checked the quantity received against that stated on the Packing Slip, by adding the totals of the full packages together, rather than multiplying. To account for less than full packages, he added each item present, rather than subtracting the empty spaces from the total, or multiplying the full rows and discounting for the empties. Satisfied that there were no inconsistencies on the packing slip, he marked it as being correct and passed it to Customer Service. Each item was brought to the storage location assigned to its supplier, where it was added to the existing stock in its original shipping 'clamshell tray' packing. Mike updated the running total of the contents of each carton of stock as an aid to stock control. The remaining empty cartons and filling materials, e.g. bubble wrap and polystyrene-foam, were stored in another part of the warehouse to be re-used in Order Filling.

Order Filling

Customer Service provides Mike with details of customers' orders in the form of a Delivery Slip, together with the shipping documents required for dispatch. Mike is familiar with the collection pattern of the assigned courier and infers the time by which the order must be ready, which Customer Service may override. Mike is faced with a decision whether to finish the order filling work that had been interrupted, or to suspend what was in progress in favour of the newly received order. He does this by estimating the finishing time of the current order, compared to that for a 'new' order and their relative priorities, and the elapsed time before the next scheduled or specially booked 'extra' collection. He decided to continue the interrupted job. Guided by the Delivery Slip, he had retrieved the required quantity of parts from stock, noting the quantity remaining on the storage carton using subtraction, and brought them to his work bench, keeping them in their clamshell tray packing. He checked again the quantity and type of the items retrieved with that specified on the Delivery Slip and that they were required to be re-programmed to suit the computer network switch environment in which the customer expected them to function.

From a possible four options, Mike selected a read/write interface platform connected to the computer on his workbench, each of which presented a plug-in socket compatible with the device to be re-programmed. He scrolled through a range of folders on the computer to locate the desired suite of programs designed to access and modify the device manufacturer's code and enable the device for use in the required network environment specified by the customer. The selected software accessed the component's internal Erasable Programmable Read Only Memory (EPROM), and displayed it in a stream of hexadecimal characters, into which Mike inserted a string of characters unique to each device. Once Mike had made the code alteration, the software ran through a series of diagnostic tests before releasing the device, and issuing a message to the effect that the reprogramming had been successfully completed. Mike removed the device from the interface platform and replaced it in the clamshell tray, flipped over, to signify that it had been reprogrammed. This routine was repeated a number of times without incident, until the computer reported an error condition. Mike retraced his steps, first verifying the part ID Code, and carefully selecting the appropriate computer program, checking whether he had made a mistake, due to a lack of concentration. Having got the same result, i.e., "program unable to load", he ruled out operator error as a probable cause. Mike considered other possible causes and his options. He restarted the computer, and loaded similar programs to try and isolate the source of the problem whether due to a faulty device, interface platform, the computer, or a specific piece of software. Finally, having tried everything he could to resolve the problem, he passed it to 'Ro', to find a solution, and informed him that the completed order was not due to be collected until the following day. Having set the 'problem' devices to one side, pending a solution from 'Ro', Mike continued reprogramming, tentatively at first, soon returning to normal fluency. When all the devices had been reprogrammed, he flipped them over to face up in preparation for the next step i.e. re-labelling.

He noted that the order called for the items to be re-labelled in a particular style. Using a label-printing computer, he selected the appropriate size adhesive blank labels, selected the customer specified font, lined up the output in a WYSIWYG display and entered the desired number of labels. Due to a fault in the computer, the WYSIWYG displayed the 'virtual' label upside-down and back to front. He has become accustomed to this over time and does not think it necessary to have the fault corrected. He printed the exact quantity of labels to match the quantity of items to be re-labelled, and then divided them into strips to match the quantity in each row of the clamshell tray i.e., the tray comprised two rows of 10, he printed a strip of 20 labels, and then separated them into two shorter strips of 10. In this way he can check that the row has been completed correctly. If he runs out of labels before the end of the row, or if he has labels left over, he has the current row only to search for error rather than the whole package. Working in columns from top to bottom, he picked up each component, matched the serial number of the individual part shown on the old label with the serial number printed on the new label. Then, using a craft knife, took the 'old' label off, peeled a new label off the printed strip, and placed it carefully on the component, which he then placed face down in the tray to indicate that the label has been changed. He did this exactly the same way every time, so that he would know the precise point at which he can re-start the process in the event of being interrupted. On completion, he sealed the clamshell tray with adhesive tape to compensate for the closing mechanism being considered unreliable from experience.

Once all the customer's specifications had been met, Mike selected an empty carton into which to pack the order. He lined up all the items and checked them with the delivery as to type and quantity. He counted trays into the carton to double-check order completeness, and surrounded them with packing insulation material to ensure their security while in transit, taking into account that their final destination was several thousands of kilometres distant. The carton's edges were sealed with heavy adhesive tape and further secured by a plastic band applied by a banding machine. Mike then estimated the weight of the carton for the courier's information, and attached the packing slip and shipping documentation provided by Customer Services. He placed the carton in the dispatch bay in the warehouse and completed the process control document, claiming the carton as his work, in the event of a query. He had been working on the shipment for a day and a half which gives some sense of the prolonged bouts of concentration, as well as adherence to the routines necessary to minimise the risk of error.

Stock Check

Mike is required to verify the accuracy of the list of items in stock provided by Customer Service. The list is ordered in strict alphabetical sequence of the item identification code; numbers taking precedence over letters. However, the stock is stored by reference to value/manufacturer and volatility and not ID code, although these may comprise elements that identify the manufacturer thus providing a link between the computer-produced stock report and the storage location in the warehouse. Mike first described the layout of the warehouse to the researcher, emphasising that the storage plan had evolved over time to suit his personal preferences and confirmed that the procedure had not been documented. He began the stock check with the highest value, most volatile components, which he recognised by their identification codes, and because they were likely to be clustered in the same location, stored in one or many packages, each manually marked with an ID code and a running total of its contents.

Starting with the top shelf of the cabinet in which the highest value items were stored, he removed all cartons and packages said to contain items bearing the same ID code and placed them on the nearby dispatch table. He checked the running total on the outside of the storage carton(s) and compared it with the quantity reported to be in stock. It did not seem to matter whether these quantities tallied, but he seemed gratified when they did. Nevertheless, he counted the content of each clamshell tray, progressively adding the total of full trays, then adding the quantity in full rows in incomplete trays, and finally adding each item in incomplete rows, in preference to multiplying the number of trays by their capacity and allowing for the partially or wholly empty rows.

In the event of a discrepancy, he checked whether the shortfall could be accounted for by cloned devices, or whether they were in quarantine and subject to a dispute, or in the test laboratory, or given out as free samples. When none of these strategies resolved the anomaly, he highlighted the line item for the attention of Customer Service. He repeated this routine for each successive set of cartons, updating the running total if found to be in error. As the stock check progressed, he moved from one location to the next, in decreasing value-priority, until having reached the point of the unchecked items being randomly distributed within the warehouse. At this stage, the balance of advantage lay with verifying the list to the items in stock, as distinct from tracing the items in stock to the list, as had been the case. The stock check process was complete when Mike reported the stock anomalies to Customer Service.

Detail Task Analysis

The WPM for Case Study 4, together with a study of the surrounding discourse, provided sufficient information for the researcher to compile Detailed Task Analyses of the three principal components of Mike's job viz., Goods Inward, Order Filling and Stock Check, reflecting the order in which the tasks were encountered and indicating their mathematics domain as perceived through Mathematical Eyes. Following the procedure as outlined in Case Studies 1, 2, and 3, the separate DTAs were consolidated and regrouped by reference to their mathematics domain e.g. Quantity and Number, Space and Shape, Pattern and Relationship, and Data handling and Chance, in preparation for alignment with the NFQ.

Aligning the Underpinning MKSC with the NFQ

Each DTA-Domain was then compared to the QQI/FETAC-NFQ Alignment Protocol, and interpreted as to the extent of evidence of meeting SLOs at any level, and annotated accordingly. There was substantial evidence of achievement at level 2 for Quantity and Number and Pattern and Relationship. Space and Shape and Data handling and Chance which warranted recognition at NFQ Level 1, even though SLOs at Level 2 were observed (Figure 5.10).

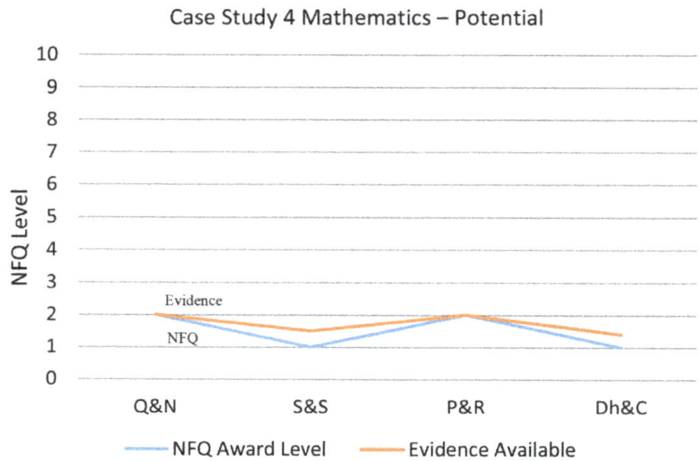

Figure 5.10. Company B, case study 4, mathematics knowledge and skills aligned with the national framework of qualifications

Implementing the rules of the NFQ in the strictest sense in Case Study 4, tends to recognise the mathematics knowledge and skills at level 1, seeming to discount the observed evidence for Space and Shape and Data handling and Chance. In the interest of clarity, by merging all mathematics domains into one, the most optimistic profile of the evidence of Mathematics Knowledge and Skills in Case Study 4 is presented in Figure 5.11, along with the complementary level of Competence in Context, Role, Learning to Learn and Insight, which is officially assumed to align with the level of Mathematics Knowledge and Skills identified.

In contrast, as shown in Figure 5.12, the researcher observed quite different levels of evidence in the Competence Strand, when interpreting the guidelines for NFQ Competence, and Chapter 2.

The standard interpretation of the NFQ provisions, in which Competence is assumed to be co-linear with the Mathematics Knowledge and Skills, assessed in terms of complicatedness, does not provide a sense of the Contexts in which they are used, even within the provisions of the existing NFQ Guidelines. Notable is that the candidate's capacity to learn about his own learning is contained by the levels of his formally recognised mathematics knowledge and skill, even though the extent of his continuous learning is indicated by the ways in which he developed his approach to his job.

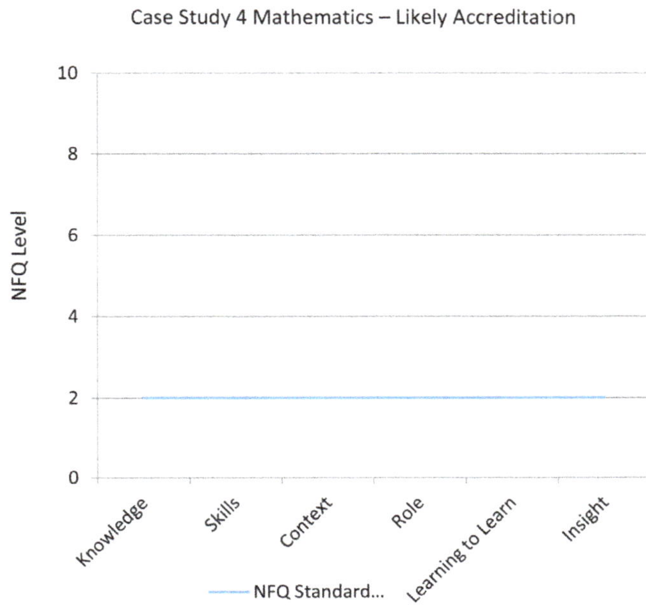

Figure 5.11. *Company B, case study 4, mathematics knowledge, skills and the competence strands of context, role, learning to learn and insight, – standard interpretation*

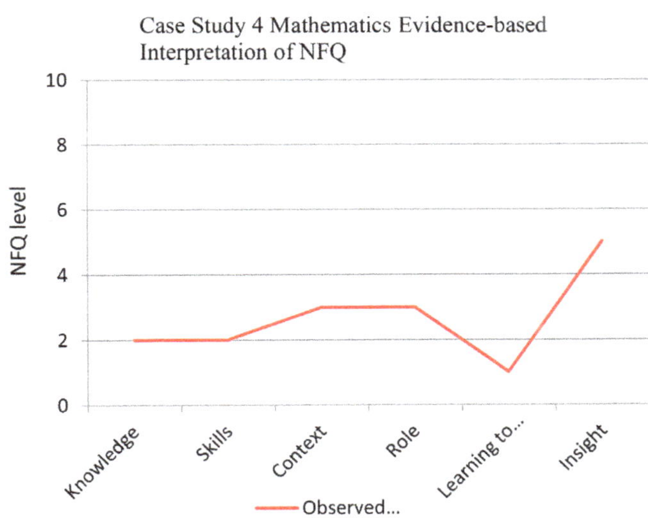

Figure 5.12. *Company B case study 4, evidence-based interpretation of the NFQ provisions to represent the underpinning MKSC*

Mathematics Invisibility: Mike

The practice of Mike's work evolved over time and in response to consequences of some sort. The relentless checking seems to be born of concern to avoid error because "*last thing you want to hear is that you got something wrong on the packing slip and it's coming back to you*". The process control document exists for two purposes, on one level to ensure completeness when two people share the work of filling and dispatching an order. At another, deeper, level it is so that responsibility for error can be pinpointed. The purchasing and logistic decisions rely, in part, on the accuracy of the stock records, which is verified by Mike, who seems to feel the burden of responsibility such that he checks the count of everything, even very low value goods that are used in their hundreds. He does not seem to trust multiplication to support the stock check, preferring instead to add in multiples of 9 and 36, and others, depending on the supplier. For example (researcher's remarks in square brackets, observations in round brackets):

> "*…9, 18, 27, 36,* [there's 9 in a row is there] *ye there's another 36, that's 72, plus 9, 10, 11 12, 13, 14, 15, 16, 17* (Not 18 minus 1, now uses calculator to add 72 and 17) *that's 89 in that box* [that's what the box says isn't it] *89, now this box*, (taking out every tray first and then counting them back into the box) *2, 3, 108, and them 2* (i.e. 36 + 36) *will be 180*". When asked about "*108*" 'Mike' said; "*because you see they're trays , they are in trays of 36 so the numbers come quickly to your head, you see a tray it's 36 if it's full* [ye] *2nd tray would be 72, 3rd would be 108* [ye] *144*".

Research discourse transcript available in original thesis. (Keogh, 2013)

Mike interprets monomial codes to locate devices, computer programmes, and when modifying hexadecimal code, all on the basis of his familiarity with these artefacts. All of his work is subject to interruption, causing him to pause in one activity in preference to another, whether to respond to an instruction, solve or escalate problem, re-start a previously stalled activity, all the while identifying, evaluating, acting upon and communicating about mathematics containing situations. That he is supported in these activities by MKSC at relatively low levels of complicatedness alone, does not capture the essence of the workplace context in which they are deployed.

Measuring Mathematics Invisibility in the Workplace

In order to contrast the self-reported numerate behaviour with that which may be observed by the researcher, each case study participant completed the survey in advance of being shadowed by the researcher. Once the job shadowing had been completed, and to minimise the researcher effect, the researcher completed the same survey to capture the numerate behaviour observed through 'Mathematical Eyes'. This enabled the comparison between the self-reported and observed numerate behaviour as indicated by the differential function of the relevant survey items, DIF. The observed Numerate Behaviour addressed Frequency items only, since the Awareness data sets are unique to the individual and not amenable to a 'second guess' by an observer. What follows is an experimental application of the DIF to examine whether there is any significant contrast in the Self-Reported and Observed data sets. This is optimistic as the sample is very small, and the expected difference likely to be < 0.5 logits. The person estimates are anchored at their values in the main analysis whereas the item difficulties are estimated for each group. This accounts for the reporting of small degrees of difference, even though the item category response is identical in some cases.

The following section takes each case study in turn and reports evidence of DIF denoting mathematics invisibility. The items displaying difference are listed in Tables 5.1–5.6, reporting their impact on any Numerate Behaviour in encounter with specific mathematics Domains, followed by specific Numerate Behaviours in encounter with any Mathematics Domain. A graphic of the estimated DIF is provided, followed by a Bond and Fox Bubble chart (or pathway) which places the Self-Reported and Observed levels of Numerate Behaviour along a scale calibrated in logits, in Figures 5.13–5.21.

CASE STUDY 1 ROB

Instances of mathematics invisibility were indicated by the DIF between Rob's completed survey self-reporting his perception of his own Numerate Behaviour (SR), and that observed by the researcher while job-shadowing. This enabled the authors to identify the underlying mathematics knowledge skills and competence, and calibrate them by reference to the National Framework of Qualifications. The analysis of the accompanying discourse revealed the subtle instances of mathematics activity that were otherwise disguised. Thus informed, a second survey to capture the researcher-observed Numerate Behaviour (OBS) during each case study in order to capture and measure the extent of Mathematics Invisibility experienced by the case study participant was completed. The findings are reported in the following paragraphs and illustrations, with a particular focus on where the self-reported (S) scores contrast with the Observed (O) scores. The resultant Differential Functioning overall profile is illustrated in Figure 5.13.

The reader will recall that survey respondents were invited to score their perception of their own Numerate Behaviour using a frequency scale ranging from 1, indicating never or hardly ever, to a high score of 5, meaning every day. The majority of Self-reported (S) and Observed (O) scores concur. Table 5.1 details and explains items where these scores differed, regarding Any instance of Numerate Behaviour in encounter with a specific Mathematics Domain.

The contrasting view of specific Numerate Behaviour in encounter with any Mathematics Domain is treated similarly in Table 5.2.

Frequency items included in the Comparison of Self-Reported and Observed Numerate Behaviour – relevant survey questions.

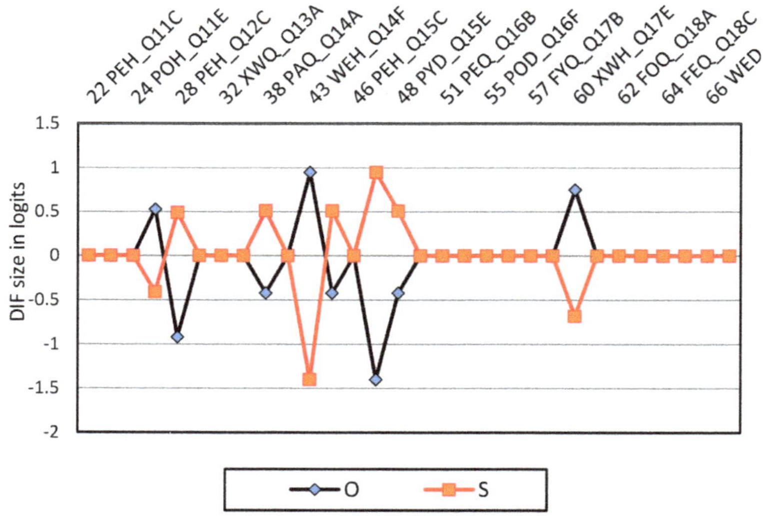

Figure 5.13. DIF size estimates for case study 1, contrasting the observed and self-reported numerate behaviours (the labels O and S are software controlled)

Table 5.1. Any numerate behaviour in encounter with a specific mathematics domain – 'Rob'

Survey Item	*SR*	*OBS*
Q15d. I merge work data from different sources to create an overall report. It was clear from the author's observation, that 'Rob' received his instructions in 3 formats, by reference to storage location, customer and delivery route. He also used Exception reports to help in the search mechanism. Finally he had to report to the line manager regarding the status of the work. It seems that 'R' interpreted this item differently hence the contrasting scores.	1	5
Q15e. I solve a problem by exploring the data for a likely explanation. The difference in score is slight, while still recognising his reliance on data handling and Chance.	4	5
Q14a. I work with amounts and quantities to solve a problem Slight difference in the recognition of frequency, however, the reconciliation of pick orders to files actually retrieved is an essential part of his job.	4	5

Table 5.2. Specific numerate behaviour in encounter with any mathematics domain – 'Rob'

Survey item	*SR*	*OBS*
Q15b. I get work information in the form of measurements and proportions Slight under-reporting on Rob's part	4	5
Q15e. I solve a problem by exploring the data for a likely explanation Slight under-reporting on Rob's part	4	5
See Table 3.15 Evaluate for overall scores		
Q12b. I measure dimensions in work The author did not observe 'Rob' measure in the course of the job shadowing. He may have interpreted this item as enquiring about matching space available for the goods to be stored.	3	1
Q12c. I notice a problem in work when the measurements and dimensions seem to be out of line with the usual. 'Rob' has become familiar with the customers and the goods that they store. He used the type size, weight and shape of the goods in the picking and reconciliation process – which is a frequent and daily occurrence.	3	5
Q14f. I read charts, graphs, sketches and the like to get information about my work There was only one instance of a graph convey information about customer satisfaction in the warehouse, which he said he did not know what it was about, other than there were testimonials from satisfied customers.	5	1
Q14a. I work with amounts and quantities to solve a problem. A frequent occurrence in reconciliation	4	5
Q15d. I merge work data from different sources to create an overall report Treated earlier in this table under heading 5.1	1	5
Q17e. I measure things The author did not observe Rob measure in the course of the job shadowing. He may have interpreted this item as enquiring about matching space available for the goods to be stored.	4	1

The revised Differential Item Functioning (DIF) profile for Case Study 1 is shown in Figure 5.14, highlighting only those DIFs that exceed 0.5 logits.

Thus, Case Study 1 offers confirmation that the survey instrument performed effectively in reflecting the individual's self-reported Numerate Behaviour, albeit with a few explainable differences. These differences can be represented as 'comparative' Mathematics Invisibility (Figure 5.15).

The larger circles depict the difficulty of the items along the logit scale. The size of the circle suggests the associated measurement error. The smaller circles, credit

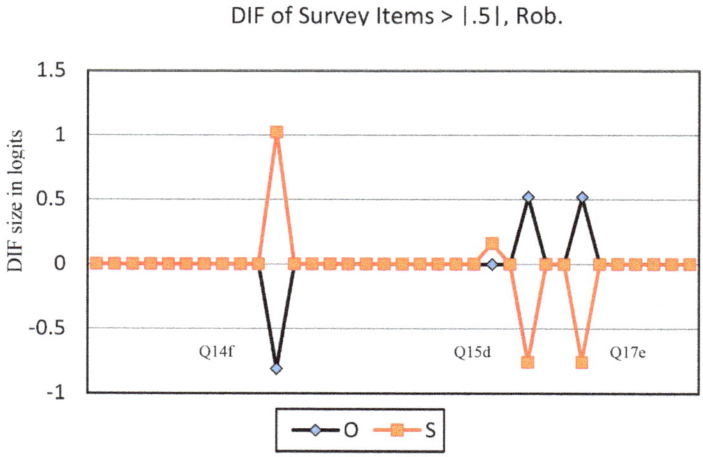

Figure 5.14. Case study 1 – highlighting the larger extent of differential item functioning only

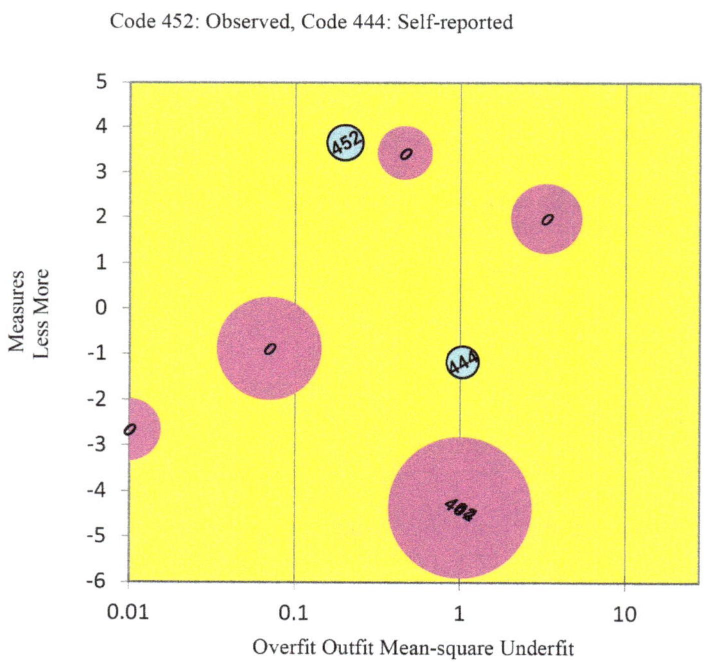

Figure 5.15. Bubble chart of the relative locations of the observed (452) and the self-reported (444) numerate behaviour

Case Study 1 (452) with more of the latent trait than he self-reports (444), especially in the items that he found more difficulty in agreeing with, e.g., he is credited with instances of Numerate Behaviour every day, instead of the less frequently.

CASE STUDY 2 JIM

The Self-Reported and Observed response sets are almost identical implying that Jim is acutely aware of his use of mathematics as evidenced by Rasch Winsteps software not being able to contrast them by degree of Differential Functioning. Although describing himself as being 'only a van driver', Jim is a very capable mathematics-competent person. The two data sets in comparison support the thrust of the survey in that workers can recognise their Numerate Behaviour when asked about it specifically. The authors formed the impression over the course of the case study that Jim was very confident in his abilities.

CASE STUDY 3 LAURA

There is a very strong coherence in Laura's responses to the survey compared to the researcher's observations, except for question 12c regarding the role she perceived for Space and Shape in her work (Figure 5.16).

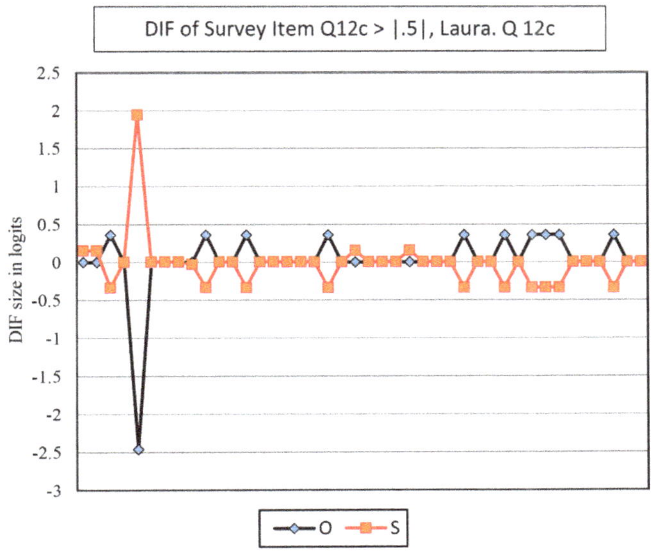

Figure 5.16. DIF size estimates for case study 3, contrasting the observed and self-reported numerate behaviours

The extent of the contrast in Question 12c influences the DIF reported in other items, even though their Observed and Self-Reported scores are identical. This view, of any Numerate Behaviour in encounter with specific Mathematics Domains is shown in Table 5.3.

Conversely, the treatment of Specific Numerate Behaviours with any Mathematics Domain is shown in Table 5.4.

It is a clear implication of Laura's response to this item, that she does not recognise her use of Space and Shape, but this is at odds with that observed by the researcher. In every other domain, Laura's perception of the role of mathematics in her job is congruent with the observed numerate behaviour. Laura is largely self-taught regarding her profound understanding of the transactional nature of her work, the need to verify, track, make projections and to optimise her resources. Her deep knowledge

Table 5.3. Any numerate behaviour in encounter with a specific mathematics domain – 'Laura'

Survey item	*SR*	*OBS*
Q13e. I convert work information from one measuring system to another	1	1
Q11e. I use measurements to describe the solution to a work problem	1	1
Q12c. I notice a problem in work when the measurements and dimensions seem to be out of line with the usual This item refers to Laura's familiarity with the suppliers packaging methods, so that, she can notice from the part number said to be included in a package, what type of carton it should be. For example antennas are delivered in 2 sizes, one being twice that of the other. Her response was that she did not understand the question.	6	2
Q13b. I think about the shapes involved in a problem to find the solution	1	1
Q14f. I read charts, graphs, sketches and the like to get information about my work	1	1
Q16d. I use the shapes involved in a problem when trying to find a solution	1	1
Q17a. I use diagrams, sketches, models and the like when I communicate about work	1	1
Q17c. I use reasoning skills on the shapes involved in a work problem when trying to figure out a solution	1	1
Q17e. I measure things	1	1
Q17f. I use the quantities of a problem to help draw a sketch, diagram or the like when discussing a problem	1	1
Q18d. I use my knowledge of shapes when trying to find a solution to a work problem	1	1

Table 5.4. Specific numerate behaviour in encounter with any mathematics domain – 'Laura'

Survey items	*SR*	*OBS*
Q12c. I notice a problem in work when the measurements and dimensions seem to be out of line with the usual	6/1	2
Q13b. I think about the shapes involved in a problem to find the solution	1	1
Q14f. I read charts, graphs, sketches and the like to get information about my work	1	1
Q17c. I use reasoning skills on the shapes involved in a work problem when trying to figure out a solution	1	1
Q18d. I use my knowledge of shapes when trying to find a solution to a work problem	1	1
Q13e. I convert work information from one measuring system to another	1	1
Q16d. I use the shapes involved in a problem when trying to find a solution	1	1
Q17a. I use diagrams, sketches, models and the like when I communicate about work	1	1
Q11e. I use measurements to describe the solution to a work problem	1	1
Q17e. I measure things	1	1
Q17e. I measure things	1	1
Relevant to Problem Solving behaviours		
No Measureable points of difference		

of the customers' patterns of behaviour together with the fluid trends in demand for a broad range of products informs her decision making, while simultaneously being conscious of the liquidity of the business as a function of credit control and remaining in good standing with creditors. The depth of her competence and the breadth of her role in the company is recognised on a personal basis, however, in describing her job a 'customer service' she renders much of her non-formal and informal learning invisible and 'out of reach' should she want to change her employment.

The contrast between the observed and self-reported Numerate Behavicur, approximately 2 logits (Figure 5.17) is much narrower that that reported in Case Study 1 (Figure 5.14), and may be illuminated by the contrast between the person's ability and item difficulty (Figure 5.18).

With the exception of Space and Shape-related Q12c, indicated by the small circle in Figure 5.18, Laura is highly aware of her Numerate Behaviour and her capacity to respond to, problem solve manage, mathematics containing situations.

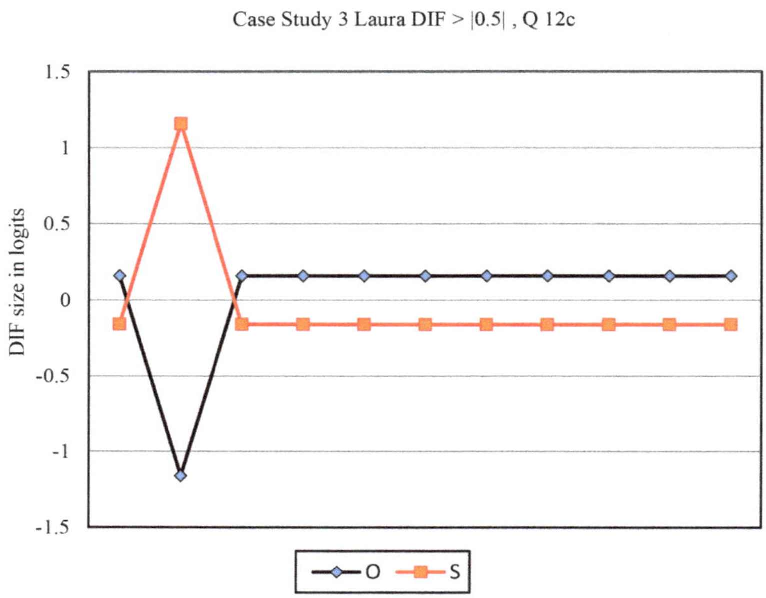

Figure 5.17. Case study 3 – highlighting larger DIFs only

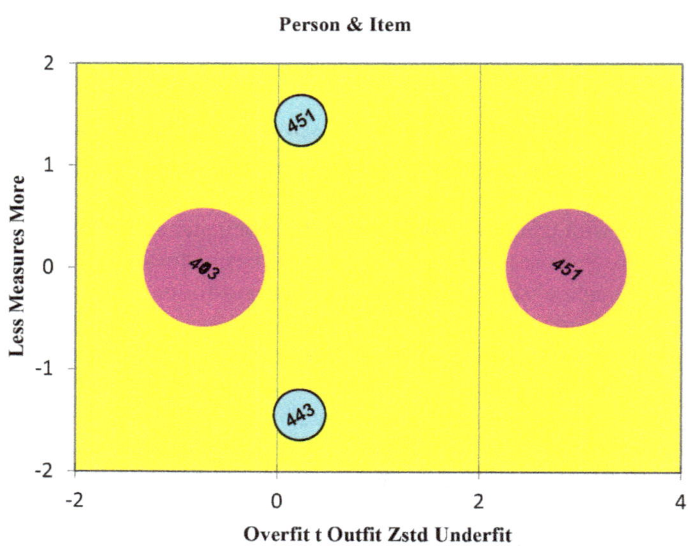

Figure 5.18. Bubble chart of the relative locations of the observed (451) and the self-reported (443) numerate behaviour

CASE STUDY 4 MIKE

The Differential Item Functioning of the Self-Reported and the Observed data sets relevant to Case 4, are extensive and contradictory, illustrating different understandings of the survey items' intentions (Figure 5.19).

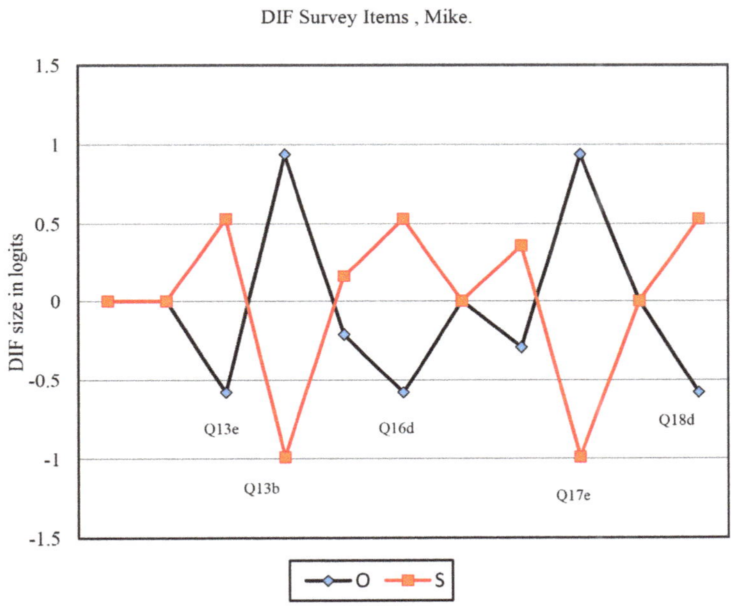

Figure 5.19. DIF size estimates for case study 4, contrasting the observed and self-reported numerate behaviours

Each affected item is taken in turn and the contrasting responses explained where possible (Tables 5.5 and 5.6), from the perspective of any Numerate Behaviour with specific Mathematics Domains.

The alternative view, i.e. specific Numerate Behaviours in encounter with any Mathematics Domain is provided in Table 5.6 and Figure 5.20.

The incidence of over-reporting frequency in response to mathematics containing situations, and under-report in others, has suppressed the contrast between the Self-Reported and Observed Numerate Behaviours (Figure 5.21).

Mike is very experienced and competent in his job. He carries out his duties with accuracy and precision and is a highly valued member of staff. He seems to be unaware, largely, of the range of his knowledge, skills and competence, even though they underpin the majority of his activities. He would be an ideal candidate to revisit, post research, to explore whether a new awareness of what he 'knows' as distinct from what he 'does' would cause him to review his self-perception.

Table 5.5. Any numerate behaviour in encounter with a specific mathematics domain – 'Mike'

Survey Items	*SR*	*OBS*
Q13e. I convert work information from one measuring system to another The author did not observe this activity in several visits to the site	5	1
Q13b. I think about the shapes involved in a problem to find the solution 'Mike' uses the shapes of items, along with other information, as clues to the solution to a problem. Components vary in size and shape, he places items in a particular way to keep track of the progress of work, he used the layout to verify the correct interface platform, find the relevant program, fit the part number to the proper label and so on.	1	5
Q14f. I read charts, graphs, sketches and the like to get information about my work. 'Mike' accesses different ranges of software matched to the products that are to be recoded, which are contained in reference lists.	3	5
Q16d. I use the shapes involved in a problem when trying to find a solution 'Mike' uses the shapes of items, along with other information, as clues to the solution to a problem. Components vary in size and shape, he places items in a particular way to keep track of the progress of work, he used the layout to verify the correct interface platform, find the relevant program, fit the part number to the proper label and so on. When doing the stock take or filling an order, 'M' locates the goods and the appropriate packing by reference to shapes and sizes.	1	5
Q17c. I use reasoning skills on the shapes involved in a work problem when trying to figure out a solution 'Mike' uses the shape of the goods, their packing, labelling and storage location every day in the performance of his work. The author suspects that these shape questions are interpreted as referring to regular geometric shapes as might be found in the mathematics classroom	1	5
Q17e. I measure things The authors did not observe this activity, although, 'Mike' did estimate the weight of a package in order to state the price point for the courier company.	5	1
Q18d. I use my knowledge of shapes when trying to find a solution to a work problem. This is a similar issue as addressed in item Q17c.	1	5

MATHEMATICS INVISIBILITY IN THE WORKPLACE

The survey findings presented in Chapter 4 deepen the Mathematics use/denial paradox. In response to a general enquiry regarding key job skills, mathematics activity was cited less than 4% of the time. That this frequency was more than 3 times that for literacy gave an early indication that the role of mathematics in work

Table 5.6. Specific numerate behaviour in encounter with any mathematics domain – 'Mike'

Survey items	*SR*	*OBS*
Q13b. I think about the shapes involved in a problem to find the solution Already addressed	1	5
Q14f. I read charts, graphs, sketches and the like to get information about my work Already addressed	3	5
Q17c. I use reasoning skills on the shapes involved in a work problem when trying to figure out a solution Already addressed	1	4
Q18d. I use my knowledge of shapes when trying to find a solution to a work problem Already addressed	1	5
Q13e. I convert work information from one measuring system to another The researcher did not observe this activity	5	1
Q16d. I use the shapes involved in a problem when trying to find a solution Already addressed	1	5
Q17e. I measure things The researcher did not observe this activity Relevant to Problem Solving behaviours No Measurable points of difference	5	1

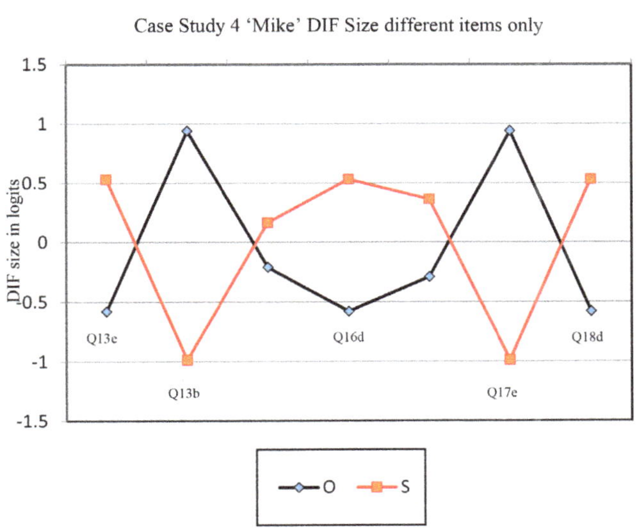

Figure 5.20. Case study 3 – highlighting larger DIFs only

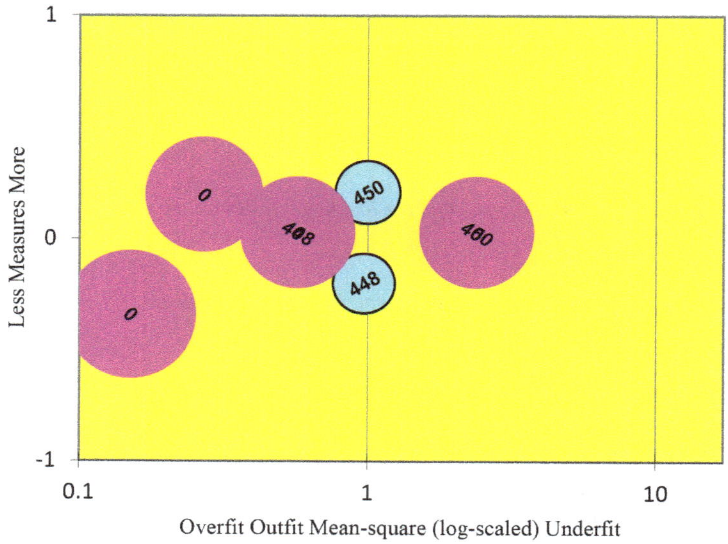

Figure 5.21. Bubble chart of the relative locations of the observed (450) and the self-reported (448) numerate behaviour

may be subliminal, in the sense of being ever present, but not in the forefront of consciousness. Weight was added to this insight by the high levels of agreement regarding the utility and relevance of mathematics concepts and their self-reported high frequency of use, in response to items enquiring about specific Numerate Behaviours in encounter with specific Mathematics Domains. The tension between these two well supported, but apparently contradictory, positions regarding human activity, prompted the author to examine the workplace through the lens of Cultural Historical Activity Theory for a coherent explanation, provided in the original thesis and in a companion document currently in press.

The authors struggled throughout the case studies activity to develop a mechanism to deal with the notion of mathematics invisibility. Initially, it seemed that there was a consciousness of the role of mathematics in work, but that it might be concealed by the way in which it was talked about. On reflection, this was not a sustainable basis upon which to begin to develop an hypothesis. To expend resources documenting the extent to which 'language' of mathematics was properly used, could only produce evidence of the non-use or misuse of appropriate mathematical terms, and would contribute little to an understanding of 'Mathematics Invisibility' or how it might be measured.

That the repetition of processes and procedures may reinforce the 'common sense' perception of the performance of work, suggested that the key to identifying otherwise invisible mathematics activity may reside the worker's Numerate Behaviour, however unconscious that might be. There were many hundreds of instances of Numerate Behaviour detected in an analysis of the discourse transcripts.

However, the key factors that emerged from the case studies, and their accompanying discourses, were the numerous ways in which Numerate Behaviour in work was influenced. The idea of the workplace being a complex, multifaceted environment, wherein the individual was compelled to juggle a number of competing demands, to perform tasks, supported by MKSC in different domains simultaneously, began to indicate that the mathematics were neither invisible nor indistinguishable, but instead subsumed into a fabric of responses, conditioned in multiple ways, to be drawn upon on demand.

The insight offered here is that if a job may consist of multiple actions and automatic operations, many of which may be performed unconsciously, the underpinning Knowledge Skills and Competence, including Mathematics, become associated with all of them as part of an automatic operation, and none of them exclusively. It is that mathematics activity is indelibly marked by the mathematics classroom only, and as such, an invisible constituent of an amorphous activity viz. the performance of work. Herein lies an explanation of the mathematics use/denial paradox. While Cultural Historical Activity Theory (CHAT) is a useful tool for disambiguating human activity systems, the authors offer a Subject-centric extension to CHAT to take account of the range of workplace activity revealed in the case study analysis. This offers an explanation for the paradox that sets a tension between the observed role of mathematics in work while being denied or overlooked by the worker.

Traditional interpretations of Activity Theory systems places activities as whole entities engaging in the pursuit of objects and outcomes which may or may not be congruent entirely. These case studies reveal a different perspective. In every case, the individual was involved in a main activity which itself was located among a palette of ancillary activities which command the attention of the individual in response

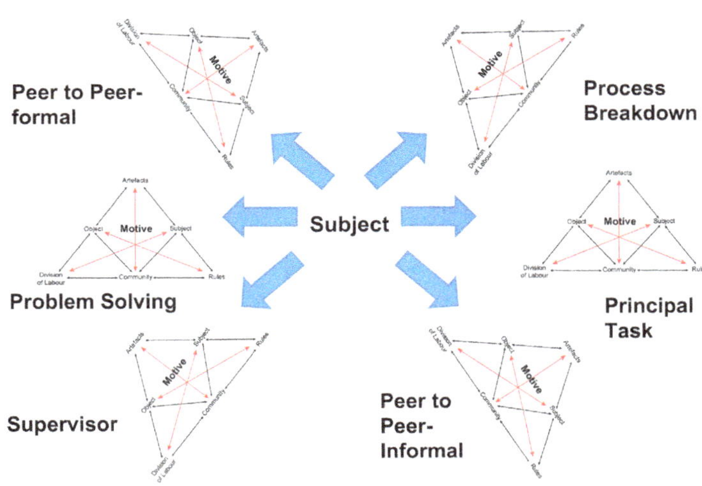

Figure 5.22. Subject-centric perspective of CHAT

to interruption or system breakdown or unanticipated interventions. For example, Jim suspended conversation temporarily while navigating a busy intersection, Laura was interrupted to answer phone calls, constantly, Mike had to deal with a malfunctioning component of the software, and Rob had to deviate from his task to locate a file that was not in its recorded storage bay. Each of these interruptions switched the workers' attention to a different set of work practices, each with its own characters. Attention was restored to the main activity once the 'interruption' has been resolved. This view of work practice places the worker at the centre of a range of complimentary activities that are interleaved and reactive (Figure 5.22), each underpinned by the same mathematics knowledge, skills and competences. This 'Subject-centric' Cultural Historical Activity Theory perspective of the work requires the worker to switch seamlessly between several activity systems, adapting and readapting continuously and without pause.

In this way, the extent of the worker's knowing becomes amorphous and invisible at the service of work practice, whereas mathematics is explicitly and indelibly linked to the mathematics classroom, thus feeding the use/denial paradox.

SUMMARY

The application of Rasch Measurement Theory (RMT) was designed to triangulate the findings arrived at by aggregating response category proportions, but it achieved more than that. It isolated survey items that did not contribute to the information-bearing capacity of the survey. It identified embedded sub-strands of opinion, and provided a platform upon which Mathematics Invisibility could be measured for the first time, adding substantially to the value of this present work.

The experience of mathematics in the workplace varies from one individual to the next. What challenges one may be such a matter of routine as to become subsumed into a matrix of other knowledge, skills and competence that are drawn on as each situation demands. That these are cognitive, reactive and perhaps reflexive events, presents particular challenges that may only be explored by an instrument that enquires about opinion using Likert style statements. The alternative approach, viz., setting mathematics tests, may succeed in discovering the extent to which individuals recall their mathematics class curriculum, and as such, have very little relevance to the mathematics in use at work. The traditional approach to dealing with opinion is to collate proportions of response categories. The assigning of numbers to enable parametric testing is convenient, but, may succeed in imposing relative value on categories of opinion that are merely ordinal. The authors were concerned that reporting proportions only may conceal underlying trends and operate to undervalue the minority, dissonant view.

RMT further served throughout Item Development, Pilot Study, the selection of the final pool of survey items, and the treatment of the survey responses. Survey items that enabled multiple interpretations were revealed and eliminated for their inability to measure the trait for which they were designed. While this approach to non-performing items may be deemed to breach the sanctity of data by discarding 'troublesome' responses, was weighed carefully. That this research did not set out to model instances of Numerate Behaviour, but rather to measure it, shifted the emphasis to those items that would measure with fidelity. In this way, it could be established whether the survey was contributing to its inferential integrity, instead of attempting to model Numerate Behaviours constructed with possibly flawed metrics. In addition, RMT enabled, for the first time, a method for measuring Mathematics Invisibility, by the transformation of ordinal data into probabilistic logit scale with which the self-reported could be compared with the observed.

Aside from the interpretation of the items addressing Space and Shape, and a few other explainable items, the findings of the survey instrument seem to have been corroborated by the Rasch Analysis of the case studies. Had the contrasts been greater, it may have been possible to isolate each Mathematics Domain and report the extent of Mathematics Invisibility experienced by the case study participant, in comparison with the authors' mathematics-sensitised eyes. Nevertheless, the evidence is strong that while people may dismiss their knowledge of-, and reliance on-, mathematics knowledge and skills in the workplace, they recognise its utility and frequency when asked about it specifically, and in a familiar discourse. It may be that the Rasch Analysis approach to the field of mathematics education has a great deal to offer in the service of future research.

DISCUSSION

To the outsider, work activity can seem to be chaotic and unstructured, surrounded by an unfamiliar codified discourse and aimed towards the achievement of an objective that is difficult to determine. In contrast, the evidence from the survey is very clear that workers adhere to procedures that are designed to cater for almost every work situation. The challenge for research in this paradigm is to enable the underlying discipline of process and procedure to become visible to the outsider, such that the layers that obscure the underlying MKSC may be systematically and consistently peeled away to make the inner core visible.

The research tools were designed to be applied in sequence to get to the core of what the worker knew about mathematics as evidenced by the rationale of what they did and to what purpose. The BAM sketched the landscape in which a suite of activities interacted to achieve business objectives. The WPM drew a fine grain focus on the actions and reactions of the worker in pursuit of a range of aims, informed by the output of surrounding activities. The DTA broke down the work practice into atomic units of knowledge that was sufficient in detail to be linked to the Significant Learning Outcomes as recognised in different levels of the NFQ. The QQI-NFQ Alignment Protocol, permitted the revealed mathematics to be mapped on to the National Framework of Qualifications. While these outcomes meet the objectives of the initiating research questions, other matters that impinge on the expression of mathematics in the workplace were drawn into sharper focus. In other words, it is not sufficient to characterise workplace mathematics knowledge, skills and competence in terms of complicatedness alone, in isolation of the circumstances in which they are deployed.

The case study selection criteria determined that the mathematics knowledge and skills uncovered would be likely to be of a relatively low order as recognised by the QQI-NFQ. The standard implementation of the QQI-NFQ is that the same level of accreditation for knowledge and skills extends to, and encompasses, the strand of competence in context, role, learning to learn and insight. There are several problems here. It is clear that the knowledge and skills that underpin the practice of work evolve to meet the outcomes required by the job without taking account of the SLOs developed to meet the needs of the National Standards. This means that the MKSC underpinning a job could range across mathematics domains and levels of difficulty, without satisfying the SLOs for any single level. In contrast, the remit of the QQI-NFQ is to maintain standards, and testify to complete mastery of the knowledge and skills of one level, indicated by achievement in a range of SLOs, before progressing to the next level.

Furthermore, the QQI-NFQ, as a matter of custom and practice extends the level of knowledge and skills assessed at any given level to the strand of competence, on the assumption that it is embedded in the curriculum and SLOs. While this may be demonstrated in the formal learning environment, its relevance to the workplace is less certain. While the level of mathematics knowledge and skills reflected in the workplace is not contested here, the application of the QQI-NFQ Competence guidelines suggests that demonstrations of competence in Context, Role, Learning to Learn and Insight in the workplace, may be interpreted as being at quite different levels to that implied by the levels of knowledge and skills demonstrated.

The QQI-NFQ dependence on the formal assessment of knowledge and skills alone, as reliable indicators of competence casts its measurement in degrees of complicatedness or difficulty to understand and master. The case studies suggest that mathematics knowledge and skills, as well as other subject disciplines, underpin the performance of work in contexts influenced by many factors in addition to complicatedness.

The workplaces featured in this series of case studies revealed that they were unique, populated by unique individuals, interacting in unique ways, each with unique outcomes. That mathematics knowledge and skills played an important part was not at issue. That the contexts in which they were used were varied and dynamic revealed a new line of enquiry regarding how such mathematics might be contextualised in the workplace. Each case study adverted to pressure of time and precision, unpredictable confluence of circumstances, accountability without authority, volatility with familiarity, the presence of distracters and competing priorities. Prominently in evidence was the constant and seamless switching of attention between multiple activities intermixed with routine, encased by constraints, restricted in scope for judgement and planning, many or all of which being conditioned by the fine balance between job satisfaction and continued employment. This observation is not offered as evidence of simultaneous multi-tasking, but rather as an indication of the continuous requirement to switch concentration between multiple interleaved activities, each with their own attributes in order to perform in a workplace featuring multiple strands of interdependent activities. These factors, and others, contribute to workplace context-complexity and is beyond the current scope of the QQI-NFQ. Nevertheless, the emergent workplace contextualisation of mathematics and the extent of its complexity offers a mechanism through which these research outcomes can be introduced to the field of adults, mathematics and their workplaces to support lifelong learning and their capacity to embrace change by repurposing their accumulated knowledge, skills and competences.

A WORKPLACE CONTEXTUALISATION OF MATHEMATICS

Strong evidence has been presented herein to show that although the procedures proliferate in the workplace, and workers adhere to their procedures, they report their work as not being straightforward. The case studies identified hundreds of instances of Numerate Behaviour in encounter with all Mathematics Domains, but at quite a modest level. The implication is that MKSC in the workplace was not captured by identifying the level of complicatedness alone, otherwise, all jobs with the same levels of MKSC could be considered equivalent.

Further analysis of the discourse surrounding the case studies, revealed that work is a social activity, having multiple properties and facets, is performed under pressure of time and accuracy, with attendant materiality, depth, scope and peer to peer accountability, not necessarily aligned with the authority conferred by seniority or role status. These dimensions arise across a range of spectra and combinations such as differentiates the MKSC required in one job when compared to another, the novice from the expert, between what a worker 'knows' and what s/he 'does', and may not feature in the official accounts of Standard Operating Procedures (SOP). Whether learnt formally, informally and non-formally, tacitly or through analogical rationality, they are highly valued in the workplace. The descriptor 'work experience' is used as a unitary concept, implying a depth of understanding that is commensurate with the quantity of time served. However, the case studies, and the findings of the National Survey of People at Work in Ireland, provided the basis for a workplace characterisation that is rather more profound and may be described more completely in terms of its Complexity. The authors have elaborated these themes to produce a contextualisation of the workplace, in terms of Accountability, Clarity, Familiarity, Stressors and Volatility in which mathematics knowledge skills and competence, regardless of level of complicatedness, are deployed. Each of these characteristics in turn comprises constituent dimensions, described in the following sections.

ACCOUNTABILITY

The common interpretation of the term 'accountability' is having to do with taking responsibility, connoting a degree of power and control as might be associated with a supervisory or management role. The corollary is that the 'ordinary' worker, for whom there are few official levers of power, is unaccountable and completely free of responsibility. The experience of the case studies suggests otherwise i.e. that accountability is a more immediate and tangible concept, comprising a range of components, in degrees of intensity that vary from job to job, each defining part of its context. It is clear that in the workplace 'trouble rolls downhill', from the most authoritative to the least.

Audit Materiality

This facet refers to the impact of error, ranging from the negligible to the catastrophic. For example, a worker in a supermarket may use the same MKSC as a person packing parachutes. An error on the part of the former is likely to be discovered in the reconciliation process and easily corrected without loss, whereas the consequences of an error in the latter could be fatal. While this contrast is dramatic, it nevertheless highlights that workers can, by making a simple mistake, compromise the service provided by the employer and expose the organisation to embarrassment, loss of business, damage to reputation and the risk of complete failure, despite appropriate procedures and SOPs.

Decision Making

The performance of work may require decisions to be made from time to time in response to unexpected or unusual circumstances. Whether the worker is permitted or expected to make decisions, to what extent, and under what conditions, extends the remit of that worker beyond simply executing a sequence of tasks. This may be further nuanced by the influence of other stressors which may produce both formal and informal interpretations of the decision making rules or guidelines.

Initiative

It is not possible to anticipate, and provide for, every possible set of circumstances that may arise in the modern, dynamic workplace. That work is not straightforward bears this out. A worker may have complete latitude to assess a novel situation and respond accordingly, or be required to apply the SOPs to the letter. There may be a 'fuzzy' understanding of when the worker is expected to use their initiative and when not. In this case, the worker assumes responsibility for having acted *ultra vires*, adding an extra tier to the dimension of Initiative component of a job, with a possible consequence of placing their continued employment at risk.

Concreteness

This is the spectrum of work components ranging from the tangible and solid to the abstract and theoretical. It is plain that, at the lowest level, manual work e.g. digging soil, comprises elements that are fully recognisable, physically present and few, whereas, at the opposite end, some or many work components may be abstract, theoretical or imagined. The gradations in between represent the transformations from one to the other, assimilating and integrating other concepts along the way. In the central range of concreteness, a trades person may handle elements that are concrete and specific, but be expected to take other factors into account such as the appearance of the finished product and its aesthetic fit with work accomplished by other people.

Judgment

There are many types of jobs that are designed with the discipline of a simple algorithm. The worker assesses the variables and selects a prescribed course of action. From time to time, the 'variables' may come into conflict and require some intervention to resolve them and permit work to continue. Such an intervention may form part of the job specification, may be conditioned or may require knowledge and expertise from elsewhere. In this way, the exercise of judgement, in what circumstances and to what extent, adds to the fabric of the context in which MKSC are deployed in work.

Planning

The function of planning is normally associated with positions of power and authority. It is an essential and pervasive strategy to increase the likelihood of completing work on time, within budget and to the expected quality standard while minimising input costs and maximising profit margin. This describes planning as a component of context at the highest end of the spectrum. The opposite end is ascribed to jobs that have no involvement in planning, although this may not be strictly the case. The authors argue that every job contains some element of sequencing tasks with the benefit of local knowledge, keeping in mind tasks that occur later in the sequence. For example, loading goods on a truck while being conscious of the delivery sequence, i.e., last-in, first-out, improves the efficiency of the unloading. In the case of all loaded goods being unloaded in one place, delivery sequence is of no account, however, the distribution of the weight, for stability may become the priority. In this way, the planning dimension of a job may be explicitly or tacitly learnt, and may be subject to rules and guidelines that vary in specificity, in its effect on context of work.

Responsibility

In the modern era, responsibility has become synonymous with guilt and the definition of who pays compensation when something goes wrong. It is associated with high status and the power to command resources and regulate activity. The authors argue that it trickles down through the hierarchy, depositing degrees of responsibility at every identifiable level, including those at the lowest level, although they may be perceived as having no responsibility. Each worker has some degree of responsibility to his/her peers, regardless of their principal activities, to produce work on time and in line with specifications. This responsibility may be assured by the processes and procedures that condition work, and escalate from partial, i.e., within varying limits to complete and absolute, thus describing another strand in the workplace context. Each of these sub-dimensions of accountability interact in unique combinations and may be influenced by the degree to which the context is clearly perceived.

CLARITY

Clarity around the aims and objectives is a desirable workplace feature, and one that is obtained in varying degrees. It is a difficult concept to describe succinctly, as its meaning is dependent on the situation it intends to describe, particularly so in a rapidly changing workplace. At every level in an organisation, it is critical that everybody has a clear understanding of their purpose, whether in anticipation of an outcome in the near, mid or long term. The authors suggest that the extent of clarity in the workplace is a combination of the interaction of several factors viz., Distracters, Priorities, Reflectivity, Information Sources, Vision and Information Completeness.

Distracters

This refers to the likely presence of elements that may distract the worker from their purpose, or add the potential for confusion and error. Simple, tightly defined jobs, involving one or few elements would seem to be free of distracters, except perhaps boredom born of narrow, repetitive cycles. Other distracters may be explicit and easily identified and discarded. Towards the upper end, it may become more difficult to discriminate between pertinent factors and distracters that are embedded and plausible. The implication for MKSC in the workplace is the ability to identify the relevant factors and select the mathematics technique appropriate to the situation. That the MKSC may be modest in terms of complicatedness does not diminish the impact of the outcome if they have been applied to the 'wrong' variables.

Priorities

The setting of priorities is a function of the control and command structure in organisations, but not exclusively so. The achievement of overall aims, comprises reaching a number of objectives, each of which may foster the development of project 'milestones', which mark the time or budget point by which a set of tasks have been completed. In the more project-mature organisations, such milestones are agreed amongst the individuals with the relevant expertise, each of whom must juggle their local resources. It may appear that individuals at the lowest end of the priority scale have no choice, having been given a simple, clear objective. Discretion regarding priorities is not necessarily aligned with job status, especially in global enterprises that commission very specific outcomes from their plants spread across the World. To this extent, the exposure to priorities, however set, is another descriptor of workplace context.

Reflectivity

Reflective practice in industry is common, although it may be realised as project review, support and supervision, strategic planning, periodic reports, performance review, and systems and financial audits. It is pervasive and hierarchical insofar as the outcomes tend to flow upstream. It may be initiated in reaction to a costly error, to identify a systemic flaw, in which case the remedies flow downstream. Other jobs may completely avoid the opportunity to reflect not appreciating its value. Most jobs could be expected to involve some degree of either personal, or professional introspection with a view to continuous improvement regarding the metrics and methods employed in, and therefore constituting, their work.

Information Sources

The sources of work information may range from single, simple source, expressed in job specific terms at the lower end, to multiple sources in various formats, referencing concrete, abstract and theoretical data on familiar and unfamiliar topics. It may be verbal and non-specific, requiring interpretation and locally-tuned inference. It may be deduced from dialogue or rumour, and vary in reliability. Dealing with multiple information sources would seem to describe a crucial element of any job, and could impinge on other context strands such as clarity, and accountability.

Vision

Vision, in this sense, has to do with the meaningfulness of the job to the individual. It alludes to the sense of purpose, beyond the boundaries of the job and how the output of the job integrates with surrounding activity to produce something that is whole in itself. For example, the collection of meter readings for input to a spreadsheet is a limiting experience in the absence of further explanation. Vision can be contained within the boundaries of a single job, or restricted by preceding and following activities each of which are mutually dependent. In contrast, acquiring a broad view of the organisation's aims and position within the market can influence the way in which work is done and the utility of the supporting artefacts, including MKSC.

Information Completeness

Work information is likely to be complete in circumstances that are tightly controlled and closely monitored, although not necessarily so. Incomplete or imprecise information, necessarily imports guesswork and uncertainty, however informed, and tends to increase the risk of error. At the leading edge of industrial research and development, complete information is the object being pursued. Creative and innovative activities feature aspects that are known and unknown in extent, and the recognition that there may be other unknown-unknowns, and perhaps even the unknowable. That this is a facet in the workplace that varies in impact on how work is done is another workplace context attribute. Exposure over time may contribute to the extent to which the characteristics and properties of the workplace become familiar.

FAMILIARITY

Familiarity is a gauge of what has become known as the 'comfort zone', a state in which the worker is well practiced in the performance of a sequence of tasks, in unchanging surroundings, in encounter with stable, recognised components. Beyond the 'comfort zone', lies the 'stretch zone', which presents challenges to the worker that are nonetheless within their capacity to achieve. The authors argue that an overall sense of familiarity, or otherwise, may be the product of the following parts:

Specificity

This refers to the extent to which components of a job are specific, recognised and unvarying at one extreme, in contrast with the abstract, theoretical, and widely varying at the other, with gradations in between to account for degrees of transformation from one to the other. The implications for the context in which the experience of work occurs are clear, encapsulating a factor which presents more challenges as specificity diminishes towards the abstract.

Principal Activity

The worker's principal activity adds a determining context characteristic. A single, closely defined and monitored, solitary activity has a simplifying effect on the worker's activity. In contrast, a professional person, at the leading edge of his/her discipline is likely to encounter a wide variety of familiar and unfamiliar situations, diagnose problems, develop creative solutions and implement them, in multiple interacting activities. In the interim, individuals may switch between increasingly varied activities in response to workplace demands.

Elements

Jobs may comprise a single element at the basic level, and progress through a single, unvarying sequence of tasks, to one that is moderately or extensively influenced by internal or external factors, some of which may be unfamiliar. This reflects complexity in the sense of the number of elements and ways in which the elements can be combined. As these quantities increase so too does the degree of complexity. This insight offers another factor with which to discriminate between jobs and the value of the derived work-experience.

Facets

Not to be confused with Elements, Facets, in this case, deals with the extent to which elements may be nuanced, not solely an empirical count. This connotes a capacity to detect and interpret a particular instance of an element and to act accordingly. Facets may become familiar over time, but that would not preclude the emergence of a novel occurrence, all of which conjures up an influential consideration of workplace context.

Group

Solitary activity can be challenging to those unsuited to working alone, but may be appropriate to a person not suited to working in a group. Engaging with a small group, becoming familiar over time, may present less challenges than belonging to a larger group that is mainly co-located. The ability to participate in an unfamiliar group, which may be large and partially or substantially distributed across a number of locations, implies a maturing set of knowledge skills and competence, and confidence in one's mathematics and other capabilities. Once again, the same MKSC may be at the service of a job located along the Group context scale, exerting an influence on the context in which they are used.

Routine

Following a familiar set of tasks in the same sequence, repeatedly, may be a product of the constraints imposed by procedure or a set of procedures, conditioned by internal or external factors. However, as the survey findings have shown, procedure accounts for just over half of workplace activity, the balance being evoked by unspecified factors such as this present workplace contextualisation is seeking to capture. Routine is a ubiquitous dimension in work, and is not completely positive in its implications, but is worth regarding for its descriptive qualities. However, many workplaces may differ from a range of factors that could contribute to stress experienced by workers.

STRESSORS

The uniqueness of the individual makes it impossible to be definitive about the causes and effects of stress in the workplace. The authors do not presume to comment on the possible effect of 'distress' in the workplace, but rather to introduce a range of factors that either singly or in combination, may change the experience of work, while using the same level of MKSC or other skills. The suggested factors are: Constraints, Pressure, Problem-potential range, Solutions, Sources of stress, and Structure of the workplace.

Constraints

In the unlikely event of limitless resources, constraints are imposed to optimise the output and minimise the input, in terms of time, materials and labour. Ranging from the clear and simple at one end of the spectrum, to those which are broad, imprecisely defined and inferred from internal and external conditions at the other, constraints have the potential to simplify or complexify work. The presence of a few clear and fixed constraints is characteristic of a job at the lower end of the scale, whereas, multiple, flexible, interrelated and mutually regulating constraints may add substantially to the performance of work towards a specific outcome.

Pressure

Workplace pressures come in many guises including the cultural, temporal, personal, inter-personal, professional, philosophical and political. Most common of these has to do with priority, urgency, accuracy and expectations. For example, completing a set of calculations under extreme and continuous time pressure is quite a different proposition to performing the same mathematical activity at leisure. In this way, the experience of work may be described by levels of pressure ranging from none, through loosely defined expectations, to issues of volume throughput targets, compliance, quality, accuracy, culminating in extreme pressure as may feature in cases of life-threatening emergency.

Problem – Potential Range

Simple jobs exhibit little or no potential for problems, excepting equipment breakdown. Even then, the worker may be required, or permitted only, to report the situation by triggering a call for attention. Jobs may increase in complexity in line with the number and possible range of familiar problems, through to levels of expertise needed to deal with multiple novel problems.

Solutions

Similarly the range of available responses to problem situations escalate from there being one response to all problems, through a continuum of the application of familiar solutions to familiar problems, progressing to mainly unfamiliar problems to that requiring novel responses and creative solutions to unfamiliar problems. Each of these levels of expertise, adds to the palette with which to discriminate between the experience value of different jobs, and the selection of the appropriate mathematics-based, or other response.

Sources of Stress

There may be none, few or many centres from which workplace stress may arise. They may be internal or external to which the individual is partially, moderately or broadly exposed. They may be such as can be avoided, or an integral part of the work, having a relentless and cumulative effect. Dealing with multiple sources of stress in work, is, potentially, more challenging to the individual, and may circumscribe the environment in which MKSC finds expression.

Structure

Working in a highly structured, tightly defined organisation, lends simplicity to its functions, albeit at the cost of flexibility, which itself might cause stress. Clarity concerning demarcation, rules, accountability and so on, may cause lower levels of stress. Loosely structured, disruptive, broadly defined organisations, may give rise to increased levels of stress as a result of their fluid, inherently unstable nature, which could be described in terms of volatility.

VOLATILITY

Volatility is the property of frequent, and unanticipated change. It refers to capacity to respond to sudden and new developments in the market or the customers' demands. The extent of volatility may be characterised as occurring over five transitions viz., completely stable, mainly stable, moderately unstable, mainly unstable, and totally unstable. Organisations and their embedded jobs, are subject to change with varying degrees of need and urgency, as may be profiled under the following headings:

Conditionality

The performance of work may be subject to a variety of conditions, the state of which may be determined by known or unknown, internal or external factors, themselves being influenced by other conditions. The range of affective conditions may differ in quantity and power. Other jobs may be immune to conditions, requiring the same response every time. The recognition of conditionality and the extent to which it pertains to a job, reflects the set of appropriate knowledge and skills and the competence, in the broadest sense, that it develops.

Demands

The demands on a job justify its existence insofar as it has been created to fill a perceived need. Simple jobs have afew demands that are clearly defined and relatively easily met. More complex jobs feature multiple demands that may not easily coalesce and may compete for resources. At this extreme, the worker sequences his/her activities, and may deploy innovative methods to cope. The effect of multiple, competing demands, may de-stabilise the job, to an extent that is unlikely in a job profiled by one or few demands.

Diversity

Diversity is the property of difference, rather than breadth. In the workplace, it refers to the extent of heterogeneity, and coherence of the tasks. While it makes sense to gather together mutually dependent tasks, requiring elaborations of related sets of knowledge, skills and competence, there are jobs that occupy the boundaries of other specialities enabling cooperation and communication. For example, a change-management specialist may need to communicate with engineers, accountants and computer software developers, in order to ensure cohesion and the desired outcome. In contrast, a completely homogenous workplace implies little scope for diversity that may not be accounted for otherwise.

Predictability

Complete predictability in a job engenders familiarity, stability, clarity, and the establishment of routine. Complete unpredictability adds depth to many of the other factors including stress, accountability, familiarity and the absence of clarity. The majority of jobs lie between these two poles, as evidenced by the survey findings and case studies.

Range

The breadth of components associated with a job confers the potential for complexity commensurate with its range. Single-issue jobs are more simple and straightforward when compared to those encompassing several issues distributed a broad, yet coherent, landscape.

Risk

In this sense, risk alludes to certainty of outcome and the extent to which it is confined. Jobs for which the outcome is almost certain e.g. attending a machine that cuts metal forms with a die, have quite a different character from stock-broking. The risk associated with the former has more to do with the wellbeing of the machine operator rather than whether the die will produce the expected form. The activities of the latter risks the organisation's resources in the expectation of substantial gain, while at the same time exposing them to potentially catastrophic loss. Risk may be classified as that component of a decision making process for which there is insufficient information. It may not be permanent and pervasive and may be conditioned and limited. Most jobs are located along a continuum between these extremes, exerting concomitant influence on the context in which Mathematics and other knowledge, skills and competence are used.

WORKPLACE CONTEXT-COMPLEXITY PROTOCOL

The Workplace Contextualisation of Mathematics represents a fairly comprehensive range of parameters with which to differentiate between jobs, regardless of the level of complicatedness of their mathematics knowledge skills and competence. The unique nature of each of each job may be reflected by the extent to which these parameters are present in the job specification and profile. That these workplace characteristics shaped the context in which MKSC was used, inspired the authors to develop an appropriate framework to capture the essence of the workplace viz., a Workplace Context-complexity Protocol, to enable the context in which MKSC are used in the workplace to be more fully reported.

Protocol Structure

Each of the main headings, Accountability, Clarity, Familiarity, Stressors and Volatility are listed with their attendant properties as sub-headings, Appendix 5. Each property is scaled across 5 transition states, each of which is described, and assigned a two-step scoring range to permit interpretation toward the lower or upper end of the scale. For example, The Volatility property, Predictability, may be scored at 5 or 6 to indicate that a job may feature moderate unpredictability that is more than the lower adjacent category (4) but somewhat less than would justify the next higher category (7), i.e. mainly unpredictable. This scoring system recognises that there is no empirical scale to measure these things yet, and that the boundaries are not sharp and clear cut. Nevertheless, guided by the evidence available and by working through each heading and sub heading in turn, it is possible to produce a detailed profile of the workplace context. In this way, the Workplace Contextualisation of Mathematics can be used as a protocol for profiling the Context-Complexity of a workplace. The idea is that it is possible to capture the complex circumstances in which fairly routine mathematics knowledge skills and competence are used in many workplaces, indexes the need for profound understanding, rather than more mathematics, by using a combination of tools (Table 5.7).

Table 5.7. Summary of the tools used in the original research and their operational counterparts

Tool	*Purpose*	*User*
Business Activity Model (BAM)	Models the context of the workplace to make sense of the activities in which the work practice is set.	Researcher
Work Practice Model (WPM)	Models the processes and procedures, constraints, priorities, adaptations and both implicit and reactions, whether explicit or implicit, formal or informal that shape the context in which the work actually takes place.	Researcher
Detailed Task Analysis (DTA)	Identifies the MKSC concepts that underpin the worker's practice in sufficient detail to be mapped on to the mathematics framework in Ireland.	Researcher
Survey	Identify and analyse the attitudes and beliefs regarding the role of mathematics of people in work.	Researcher
Task Exploration Tool	Operational version of the BAM.	Practitioner, Trainer
Task Context Tool	Operational version of the WPM	Practitioner, Trainer
Linking Tool	Operational version of the DTA	Practitioner, Trainer
Workplace Context Complexity Protocol	Supports the worker, trainer, teacher, curriculum developer, change manager, recruiter, and employer to highlight the range of workplace knowledge, skills and competences across a scale of intensities, that may be invisible or dismissed as 'just part of the job'.	Practitioner, Trainer

That the individual's use of mathematics in the workplace may be denied or dismissed as common sense, argues in favour of a mechanism that is capable of making the mathematics more visible and more fully accounted for. Chapter 3 provided details of the extent to which the complicatedness of mathematics may be aligned with the NFQ, but with some caveats. This present work now suggests an augmentation to the NFQ to facilitate the recognition of, and communication about, mathematics activity in work, and how the research can be operationalized for the benefit of society.

CHAPTER 6

FROM RESEARCH INTO PRACTICE

How to Bring the Benefit of Research to Learners

This chapter is devoted to considering the impact of this research on practitioners including schoolteachers, work-based trainers, workforce/career developers, policy makes etc., as they work in their respective environments. This is important in a dynamic landscape where employability and re-employability of workers is of paramount concern for themselves, their employers and society in general.

Through this research, the authors have made a significant contribution to the understanding of the reported phenomenon of mathematics invisibility in the workplace i.e. the mathematics use/denial paradox. The research also locates mathematics as a bundle of competencies that are associated with many activities generally, but only the classroom explicitly. This work offers a key to resolving the use/denial paradox, not only as an end in itself, but to help workers to recognise their knowing, to embrace it, and to be able to repurpose it to different contexts and for different outcomes.

A key objective of the research was to identify instances of mathematics in work and to map that on to a recognised official framework, in this case the National Framework of Qualifications in Ireland which is aligned with the Bologna Education principles.

The research literature shows that workers are sensitive to what they do and valorise it, very often dismissing what they know as common sense. The authors' research shows that there is another important dimension. The workplace environment for workers, is shaped by many factors that interact in complex ways, based not only on topic-specific knowledge skills and competence.

The authors posit a framework of new facets that capture and measure these influences. The framework responds to the CEDEFOP recommendations for a more systematic approach to validating non-formal and informal learning wherever it occurs.

The CEDEFOP guidelines for validating non-formal and informal learning recognise that learning throughout life is a key route to an individual's personal and professional development. National accreditation frameworks e.g., NFQ in Ireland, capture the formal educational experience, the non-formal and informal learning is often difficult to measure and is, as a consequence, undervalued.A more systematic approach to validation would increase the visibility and value of the non-formal and informal learning outside formal education and training systems. This in turn would benefit individuals by enabling them to communicate the value of their personal and professional knowledge, skills and competence to current and future employers or

 DOI:10.1163/9789004381766_006

further education and training providers. It would support employers in matching the real capacity of the workforce to their tasks and to identify and meet training needs in a more precisely targeted way.

Validation is, first, about making visible the diverse and rich learning of individuals. This learning frequently takes place outside formal education and training – at home, in the workplace or through leisure time-activities – and is frequently overlooked and ignored. Validation is, second, about attributing value to the learning of individuals, irrespective of the context in which this learning took place (Cedefop, 2015).

Furthermore, this framework enables the credentialising of knowledge, skills and competences that are newly made visible, while having been important all the while. These things were hinted at in the survey but emerged very clearly as self-supporting and consistent in the case studies, bestowing a foundation of both validity and reliability to how these insights can be re-contextualised from the research domain into practice.

That the individual's use of mathematics in the workplace may be denied or dismissed as common sense, argues in favour of a mechanism that is capable of making the mathematics more visible and more fully accounted for. This present work now suggests an augmentation to the QQI-NFQ to facilitate the recognition of, and communication about, mathematics activity in work.

EXTENDED NFQ ILLUSTRATED

The provenance and provisions of the National Framework of Qualifications was discussed extensively in Chapter 2. Its capacity to recognise mathematics activity in the workplace was discussed in Chapter 3, together with an account of the challenges posed by the dissonant structure of the workplace in comparison with that of the formal learning environment. The formal granting of a level of Competence in line with that earned by mastery of Knowledge and Skills was contrasted with the evidence of a 'spiky' profile of workplace Competence, which was often gauged to be a higher than that of the mathematics knowledge and skills in evidence. Notable too was that the 'Learning to Learn' sub-strand of Competence was predicated on being acquired in formal learning structures, with little or no recognition for non-formal, informal, tacit and other prior learning opportunities as typifies the workplace. Further, it became apparent in the course of the case studies, Chapter 5, that calibrating MKSC in the workplace by reference to mathematics complicatedness alone, did not fully account for complex contexts in which the MKSC was deployed. As a result, a Workplace Contextualisation of Mathematics, and the Context-Complexity Protocol which it underpins, is explained in detail in the original thesis and in a companion document (ICME, 2016, in press).

The close observation of four case studies highlighted the absence of formal recognition for several powerful learning opportunities and outcomes that arise in workplaces, which the NFQ, in its present form, is unable to recognise.

EXTENDED NFQ – CASE STUDY 1

The standard NFQ approach to the accreditation of learning when applied to Case Study 1, represents the identified mathematics knowledge and skills at level 1, having met the criteria detailed in the relevant Significant Learning Outcomes (SLO). In keeping with standard custom and practice, the same level is credited to the four Competence sub-strands, viz., Competence in Context, Competence in Role, Competence in Learning to Learn and Competence in Insight, shown separately (Figure 6.1).

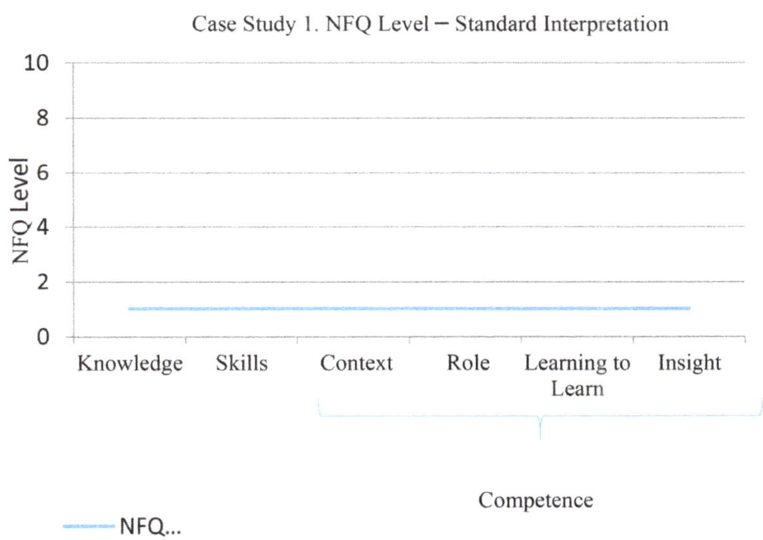

Figure 6.1. Company A, case study 1, mathematics knowledge, skills and the competence strands of context, role, learning to learn and insight, – standard interpretation

This graphic is not intended to denigrate the competence strand associated with formal recognition. Instead, it highlights that this competence strand is confined to the formal classroom, the study of the course material provided, and the successful completion of tasks designed to test understanding and recall of that material and very little else. Competence in the workplace is rather more than that.

Competence in Context and Role, based on the evidence of observations and on interpreting the formal provisions of the QQI-NFQ, exceed that of the mathematics Knowledge and Skills levels identified in the DTA (Figure 6.2).

The scale of Learning to Learn recognition is biased in favour of the formal learning structures, leaving no scope for the recognition of tacit or other learning. The apparently extreme score recorded for Insight, reflects intelligent exposure to the workplace, and its capacity to promote tacit rationality and analogical thinking.

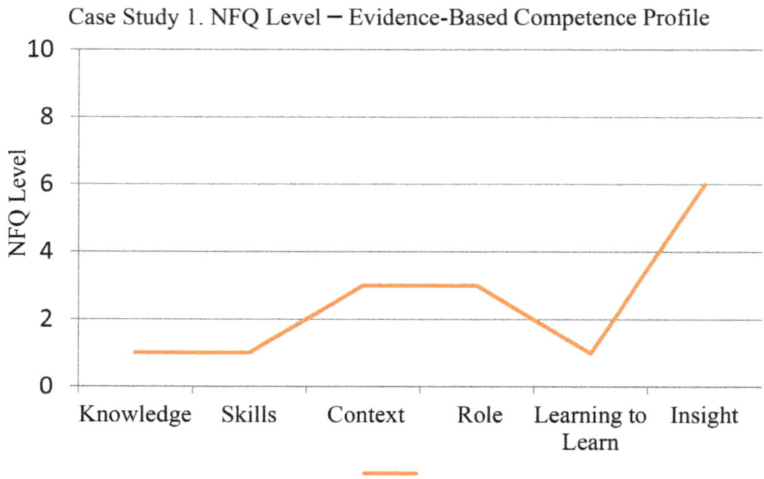

Figure 6.2. Company A case study 1, evidence-based interpretation of the NFQ provisions to represent the competence observed in the workplace

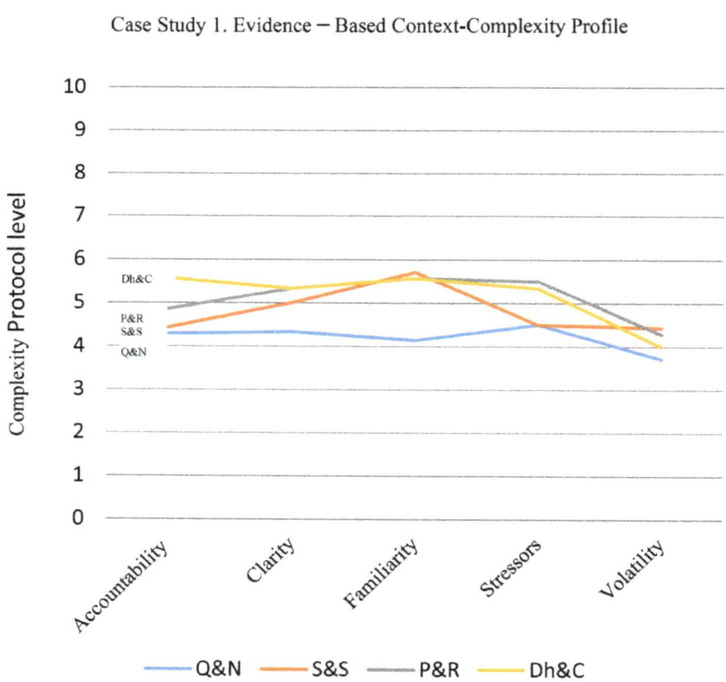

Figure 6.3. Context complexity dimensions for 4 mathematics domains

The impact of having applied the Context-Complexity protocol to Case Study 1, is shown in Figure 6.3. A separate trace is shown for each mathematics domain to maintain consistency with the underpinning Detailed Task Analysis (DTA). While each data point, for each factor is available, the plots shown represent the mean score of the factors comprising each dimension of complexity dimension.

When these data are combined, the resultant graphic (Figure 6.4), captures not only how complicated the Mathematics Knowledge and Skills deployed in this particular workplace, it also shows the observed, rather than assumed, competence in Context, Role, Learning to Learn and Insight, appropriate to each Mathematics Domain. The key additional job profile information reports Context-Complexity dimensions which are imposed by the workplace featuring dimensions of Accountability, Clarity, Familiarity, Stressors and Volatility, each comprising sub-strands as explained in a companion document (Keogh, Maguire, & O'Donnoghue, 2015), and in the original thesis (Keogh, 2013).

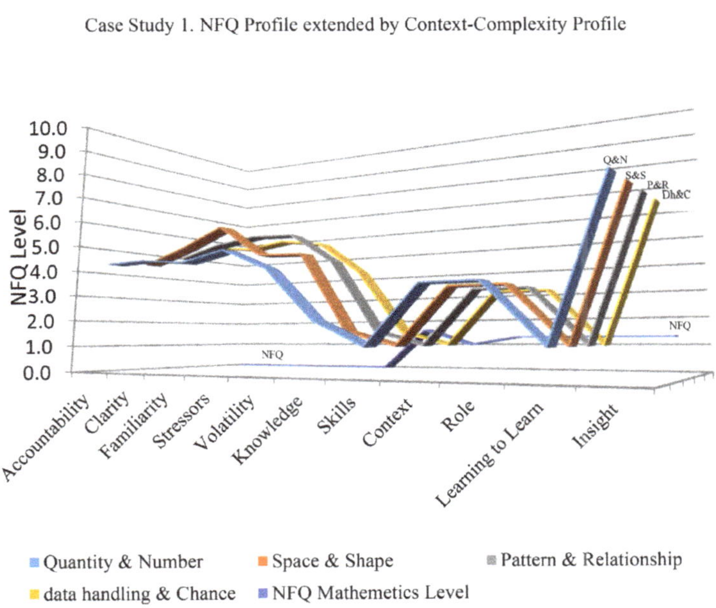

Figure 6.4. NFQ extended to account for context-complexity

In sum, MKSC identified in the workplace and aligned with the existing provisions of the NFQ, is rooted by the level of complicatedness at which it has been assessed (Figure 6.1). Allowing for an assessment of competence observed in the workplace, independently of its level of complicatedness, recognises a more sophisticated context and role that might be derived from knowledge and skills only. Insight is acknowledged as the integration of many facets of the workplace experience. However, it is the profile of the context-complexity that communicates new information about the workplace and a sense of what the individual 'knows' in addition to what s/he 'does'.

The graphics in Figures 6.5–6.8, contrast the information-bearing capacity of the standard NFQ in comparison with that provided by the Extended Model, proposed herein.

EXTENDED NFQ – CASE STUDY 2

The Mathematics deployed in case Study 2 meet all of the SLO requirements prescribed for NFQ Level 1, as determined by the property of complicatedness, which extends across the dimensions of competence. However, observation of the work practiced in this case study, interpreted using NFQ provided guidelines, presents a more varied and 'spiky' competence profile (Figure 6.5). The level ascribed to Learning to Learn is aligned with that for Knowledge and Skills, as it is gauged from the formal learning environments perspective.

The application of the Workplace Context-Complexity Protocol to the evidence offered by DTA, and the surrounding discourse, suggests a more complete profile in an extended version of the NFQ. The NFQ profile is included for contrast and does not register levels of complexity (Figure 6.5).

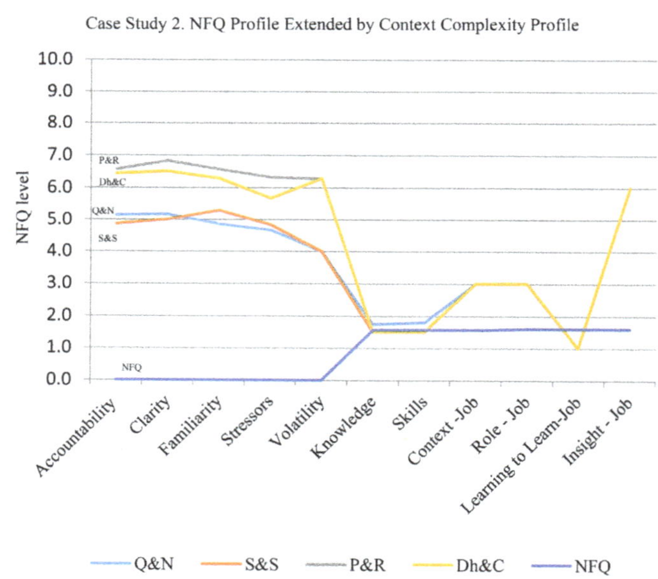

Figure 6.5. Case study 2, workplace MKSC extended to account for context-complexity (note: plots that coincide are not visible as separate entities)

EXTENDED NFQ – CASE STUDY 3

The MKSC observed in Case Study 3 warrants recognition at NFQ Level 3, as extended to the four Competence strands of Context, Role, Learning to Learn and Insight. The role occupied by this Case Study is set in a very diverse and unpredictable context, requiring high levels of leadership and social skills, normally associated with higher levels of formal learning. It is volatile, dynamic, and reactive to changing circumstances in a very competitive commercial landscape, where insight is highly valued.

While the complicatedness of the MKSC at NFQ Level 2, may be characterised as routine, the circumstances under which they are used requires their profound understanding, in support of thinking and as a guide to action. An extended profile of the competence associated with case Study 3 is shown in Figure 6.6. That the different Mathematics Domains seem to be deployed in equal measure suggests that they are used in continuous, rather than discrete ways. Noteworthy is the relatively minor role occupied by Space and Shape, which is consistent with the findings of the survey.

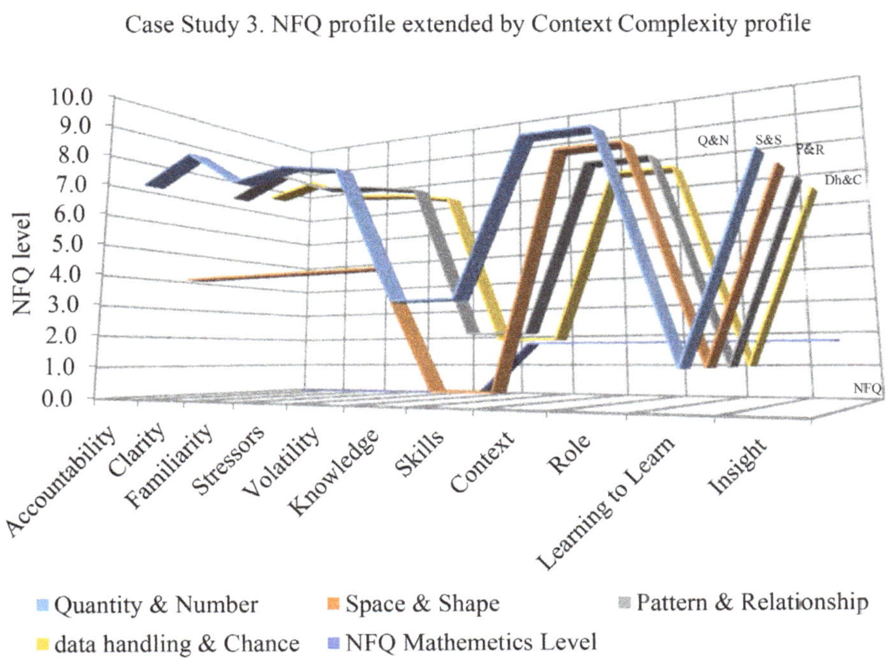

Figure 6.6. Case study 3, workplace MKSC extended to account for context-complexity (note: plotted points that coincide are not visible as separate entities)

EXTENDED NFQ – CASE STUDY 4

Although Case Study 4 is classified as being similar to Case Study 1, insofar as they are both described as 'Warehouse Operatives', the two jobs are quite different. The MKSC observed in Case Study 4 align with the NFQ at Level 2. 'Mike' services the complex environment in which 'Laura' operates and this is reflected in the extent of his Competence in Context, Role and Insight. Learning to Learn, being anchored by the formal setting requirement, does not recognise the extensive tacit learning acquired by 'Mike' in the course of his work and experience.

While not attaining the heights of Case Study 3, 'Mike' shares the complexity of the workplace and all that that implies. This is deduced from the DTA and surrounding discourse, and more fully reported in Figure 6.7.

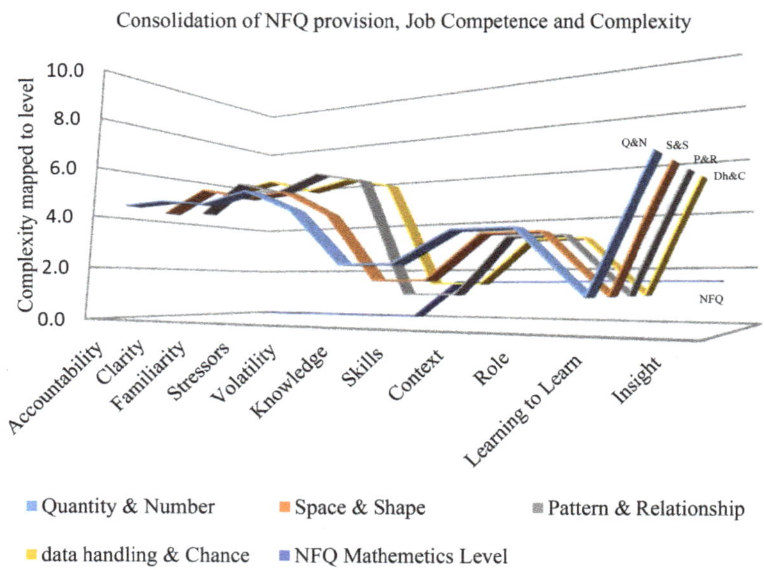

Figure 6.7. Case study 4, workplace MKSC extended to account for context-complexity (note: plots that coincide are not visible as separate entities)

RECRUITMENT DICHOTOMY

The design and implementation of the NFQ has served the formal education environment well since its inception more than a decade ago. It provides a platform for the maintenance of standards, and the recognition of awards achieved outside its jurisdiction. It testifies to the academic calibre of the individual in terms of his/her chosen subject discipline, to such an extent that it is cited as an all-encompassing descriptor of the ideal job candidate in many cases. There is no doubting the standard but an opportunity is missed to recognise non formal, competency bundle that surrounds it. The following profile of the knowledge and skills attested to by the NFQ level, including the assumption of Competence being in line with the level of knowledge and skills, may need closer examination as a result of this present research (Figure 6.8).

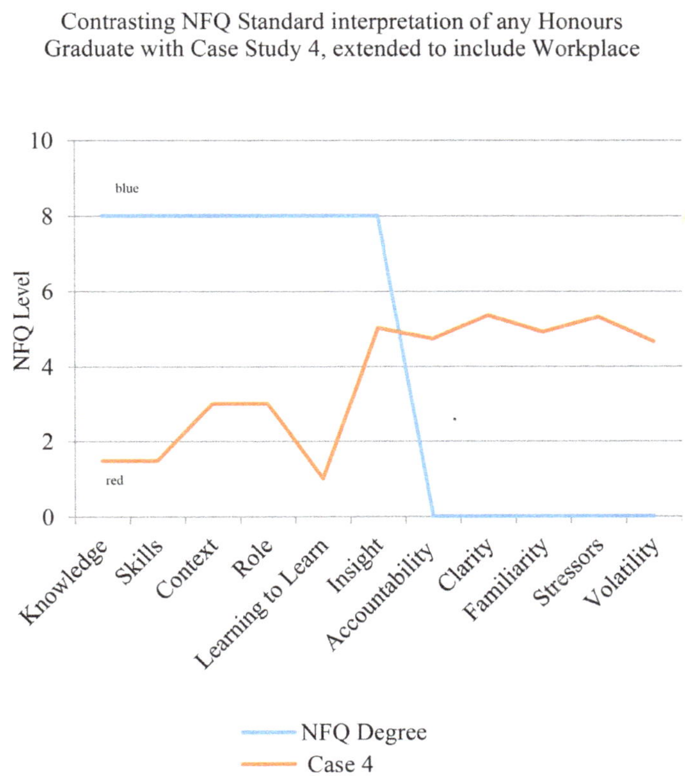

Figure 6.8. Comparison of preparedness for workplace content complexity

The blue line traces the formal accreditation awarded having successfully completed study at level 8, Honours Degree, extends to the sub-strands of competence, although they are neither explicitly taught nor assessed, but may be learnt differently by different students. However, they are contained by the classroom and topic discipline, and, while it testifies to success in the classroom setting, it offers very little indication of work-readiness. Since there is no formal provision for the recognition for context-complexity, this dimension registers zero for the graduate.

In stark contrast, the red line, representing the findings in Case Study 4, for example, begins with effective mathematics knowledge and skills in the workplace at less than level 2, similarly for Learning to Learn, due to the formal and structured learning recognition constraint. From there, noticeably higher levels of insight are recorded, accompanied by observed competence in complex contexts. It seems that while employers value competence in complex contexts, they recruit by reference to academic qualifications *alone*, thus disqualifying the candidates whose competence they seek.

This insight is not intended as a criticism of academic qualifications, but rather to highlight an anomaly which can be resolved for widespread benefit. Informed by this profile, curriculum could be modified to take account of relevant workplace context complexity. It is of particular benefit to unqualified or less qualified persons who could be encouraged to review their self-perception of 'not being a maths person', and pursue their own development by exploiting their workplace complexity-competence to help them achieve relevant qualifications. The recognition of prior acquired learning, while supported by the proposed NFQ extension and elaborated by the Contextualisation of Mathematics and implemented by the Context-Complexity Protocol, it is unlikely to serve those for whom their experience of Mathematics Invisibility persists. In other words, in order for individuals to claim competence in mathematics, they must first become aware of it themselves, by overcoming their denial of use of mathematics in work.

IMPLICATIONS: LEARNERS, CURRICULUM, TUTORS, POLICY, STANDARDS & RPL

It is clear that tension between the context in which mathematics is learned and that in which it is realised, adds to the difficulty of achieving any measure of transfer between them. It is not that the mathematics techniques are any less important in one context than another, but rather that they need to be re-contextualised to become relevant and useful. That this transition occurs by an osmosis-like process tends to render the underpinning mathematics 'invisible' to the worker or at least 'other' than that experienced in the classroom, thus compounding the self-perception of not being a 'maths-person'. There is no doubt that this transition occurs already as evidenced by the workers' capacity to adapt to their work, but this may happen with different rates of success and completeness depending on the workers' aptitude and the efficacy of induction arrangements. Successful transitions may be thought of as evidence of workplace-experience and described as common sense which tends to diminish its importance. Furthermore, the modern workplace landscape is necessarily dynamic in response to technological and regulatory change, which, increasingly, features jobs being replaced by automation, i.e. the Fourth Industrial Revolution i4.0 (Gates & Bremicker, 2017) . The received wisdom is that, similar to earlier cycles of industrial innovation, jobs made redundant by emerging technology will be replaced by new technology-related jobs with existing resources being 'released for re-deployment' all the while. While the supporting evidence has yet to be identified, such fluidity in the workforce opportunities may be ameliorated by providing a mechanism with which to identify what a worker 'knows' as distinct from what s/he 'does' complemented by a method for analysing a workplace context for which existing knowledge, skills and competence may need to be re-purposed.

The policy of 'Lifelong Learning' recognises that knowledge, skills and competence must remain in lock-step with the continuous evolution of opportunities and challenges of society in work and at leisure. While learning is not exclusively purposed for the workplace alone, its relevance to preparing people for work is a key objective nevertheless. The authors argue that:

- Curricula are designed in response to the demands that are thought to comprise a subject discipline
- The summative award associated with formal learning reflects mastery of learning outcomes that may, or may not, prepare the learner for work
- Learning a discipline may take several years, *ab initio ad finem*, during which time standards may have changed and expected learning outcomes diminished in relevance in relation to the workplace
- The existence of work-placement programmes testifies that syllabuses, realised by a range of curricula, may not constitute sufficient preparation for work
- Work-placement roles are shaped by the discretion, need and opportunity of the work-placement host
- Outcomes of work-placement programmes are inconsistent and patchy, dependent on the discipline, host, and assessment methods
- The context of the classroom in which the curriculum is learned stands in sharp contrast with the context in which it is realised
- The expectation of a stable, 'job for life' is not sustainable in the current workplace landscape
- Knowledge, skills and competence acquired outside formal education settings, though highly valued, are not formally recognised and not portable between jobs, across industry sectors and beyond national borders
- Recognition of Prior Learning is predicated on the worker's awareness that a particular competence, stripped of its contextual significance, is 'recognisable' and potentially exempted in an existing syllabus as a component of a more holistic accreditation
- Workplace competence is real, highly valued and typically dismissed as common sense that is more noticeable by its absence i.e. when something 'goes wrong'.

In this light, the expected transfer of knowledge and skills may be frustrated for the want of a channel that facilitates the re-imagining of previously acquired MKSC for new purposes. Such a channel as is capable of identifying existing knowledge by reference to a workplace context, and calibrating a new, destination context could stand on the workplace contextualisation of mathematics and the protocol that it inspired, Appendix 5.

This work suggests that a mechanism to capture workplace competence would support all stakeholders whether contributing to Policy, Syllabus, Pedagogy, work-placement, employability, re-employability and mobility within and across industry sectors, including volunteers, nationally and internationally.

RESEARCH INSTRUMENTS MODIFIED FOR PRACTICAL IMPLEMENTATION

While the research instruments were designed to meet international standards of intellectual rigour, they may be condensed to more manageable proportions that may be applied in a variety of situations. In the first instance, a more concise, operational version of the survey comprising validated items from the original, but shortened for operational convenience, may be self-administered or completed with the support of a practitioner. This helps the learner to recognise their numerate behaviour and begin to review their 'non-maths person' self-perception, Appendix 6. The essence of the BAM, and WPM is to describe and identify workplace implementations of mathematics techniques, and may be replaced with semi-structured Socratic scripts as the basis of enquiry, e.g. Task Context Tool, Appendix 7, and Task Exploration Tool, Appendix 8. Together with the guidance of the Workplace Context-complexity Protocol, Appendix 5, these instruments can be used to:

- Answer questions regarding the relevance of curriculum content to the workplace/ professional environment
- Familiarise the learner with aspects of the workplace that contrast with the mathematics classroom, and that vary in combination and intensity
- Provide a platform with which to identify the sophistication of the context in which they do their work, nuance their MKSC and disambiguate what they know from what they do
- Underpin their employability, re-employability and mobility within and across industry sectors at home and abroad.

HOW THIS APPROACH WORKS

Concise, Operational Survey

The workplace survey instrument, having been validated as previously described was reduced to more manageable proportions such that it may be self-administered or with the support of a practitioner. It is designed to prompt self-reflection on the learner's activity, whether social or professional, by locating them in the midst of their own expertise. In this way, the learner begins to appreciate the mathematics underpinning of their work and/or play and realise that the impact of mathematics ranges across space and shape, pattern and relationship, data and chance, as well as quantity and numbers. It presents a mathematics-nuanced perception of what may be seen otherwise as common sense or anything but mathematics.

Task Context Tool (Inspired by the BAM)

The purpose of the Task Context modelling tool, Appendix 7, is to build on the outcomes of the survey. It is administered by a practitioner or mentor to model the context in which the learner expresses their chosen expertise, whether in gainful employment or otherwise. It is a purposeful enquiry into what the learner does, described in his or her terms, without being led by the practitioner. However, it is a Socratic pursuit of information that might be dismissed as common sense otherwise. The expected outcome is a diagram that depicts the flow of actions and information that underpin the aims and objectives of the activity, including the attendant motivations, constraints, conditions and problems. The semi-structured script comprises questions that seek clarification, probes assumptions, reasons and evidence, elicits viewpoints and perspectives, and tries to capture implications and consequences. Suggested questions neither challenge nor judge, but intend to place the practitioner in the learner's shoes as follows:

- Tell me about your work, hobby, past time
- What happens there?
- Why?
- How do you know what to do?
- What happens next?
- How does that happen?
- Have I got this right so far?
- What have I left out?

The key output is that the learner shares the context of their work or pastime as a precursor to describing their role and how it fits into the 'big picture'. This is a cyclical process shaped by questions that expose many layers of detail that may be concealed by a summative job title, performed by a practitioner who must guard against making or imposing assumptions. The diagram of the context serves to acquaint the practitioner with the locus of the learner's area of expertise, rapidly, to ask intelligent questions and to understand the answers, to enable a closer understanding of what the job/leisure pursuit entails in terms of the tasks that are 'done', in order to access the extent and kind of the underpinning 'knowing'.

Task Exploration Tool (inspired by the WPM)

The Task Exploration Tool, Appendix 8, drills deeper into what the learner actually does in order to uncover the numerate behavior that s/he relies on. This is a more detailed enquiry of the processes and procedures that guide the work activity rather than a list of duties and is informed by the outcome of the context tool. Particular attention should be paid to codified references and acronyms that tend to conceal rather than explain. While many of the task components may be superficially repetitive, they can be constrained by sets of conditions that temper the worker's responses. It is useful to derive a diagram from the information provided so as to identify omissions and anomalies which may point to actions and decisions that have become invisible through familiarity. This work too is supported by a semi-structured script which includes the following question types:

- What is the first thing you do on arrival in work?
- Why?
- What happens next?
- How do you know what information/resources you need?
- What can go wrong?
- What solutions have you for when things go wrong?
- How can you avoid things?
- What makes things awkward?
- How would you change our work?
- What causes you trouble?
- How do you deal with it?
- What is not covered by the SOPs?
- How do you work around problems?
- What makes one day different from the next?
- What's the last thing you do?

This is not an exhaustive list, nor should it be considered a prescribed script, but a methodology for stripping back the occluding layers to access detailed numerate behavior. A detailed example is provided in Appendices 7 and 8. The information provided should be sufficient to enable the development of a diagram of the relevant actions and their conditions described in as atomic detail as possible, such that they may be numbered.

Linking Tool (Inspired by the DTA)

While this provides the practitioner with a sound basis to work from, some may find it necessary to link the underpinning mathematics to a more formal description of the domain and level using the Linking Tool, Appendix 9, which is self-explanatory and flexible enough to accommodate the level of mathematics that matches the learner's ambition. In the course of the application these tools, the practitioner is tuned to hear clues as to the numerate behavior embedded in the activities the learner describes. This enables the practitioner to establish the link to the SLOs and to demonstrate the learner's familiarity with the underlying mathematics concepts with which they have demonstrated their familiarity, albeit expressed in non-mathematics terms.

In sum, the survey prompts the recognition of numerate behaviours that may have been dismissed as common sense – something everybody knows. The Context Exploration Tool places the learner in the zone of their competence and comfort. That the practitioner asks probing questions helps the learner to 'unpack' the ease with which s/he commands the vision of his/her job and the grand scheme of things, revealing aspects that may have become invisible through familiarity. Similarly, only more deeply, the Task Exploratory Tool teases out the 'ifs', 'ands', and 'buts' that shape the detail of their numerate behaviour. All the while, the practitioner is mindful of the Linking Tool which is used to bridge the learner's numerate behaviour to the curriculum relevant to the learner, to inform their pedagogical approach to lead the learner from where s/he perceives themselves to be, to the learning they seek. The Linking Tool closes the loop between numerate behavior, its mathematical provenance and its location in the formal spectrum of mathematics domains and level that is appropriate to the learner. The depth and precision of its application depend on the aims of the practitioner and draws on his/her relevant experience.

WHY DOES IT MATTER?

There are several converging strands that compete to attest to the achievement of what are thought to be complementary outcomes. Learners have needs that are influenced by their interests, cultural background, career objectives and what they perceive to be their innate talent. Policy makers encourage and support the development of knowledge skills and competence for the benefit of society in the broader sense, while at the same time catering for the needs of industry by prescribing standards of quality and consistency. Practitioners and trainers devise curricula that comply with the terms of the national standards while at the same time appealing to and meeting the ambitions of the learners, insofar as they can be known in advance, and measured with reference to learning outcomes.

Employers operate in a more dynamic and fluid context that prioritises marketability, cost efficiency, competitiveness and profitability, but the implications of the NFQ levels are regarded in the broadest sense only. The factors that drive these things are made volatile by the necessity to react to change in demand, technology, innovation and expectation. It is impossible, despite Government funded initiatives, to anticipate fully the sets of knowledge skills and competence that may become scarce within a few years and therein lies the kernel of the problem. Educators must equip learners with the resources to perform in an environment that is undergoing constant change. This must be done in compliance with a programme that has been in gestation for two years prior to being implemented in a four year degree programme. A person in employment for that same six years will have acquired a profound understanding of the workplace and its requirements, albeit specific to and probably contained by that workplace.

Both of these learning paths are valuable in their own way, but neither is complete nor future proof. The academic route is contained by prescribed learning outcomes that are recognised as formal 'knowing' but not immediately useful in the workplace. The 'learning on the job' route aggregates usefulness that is focused on being able to 'do things', but is seldom negotiable as competence in the pursuit of re-employment, partially because of not being formally recognised, and because different workplaces 'do things' differently. This work offers a workable solution that is underpinned by the rigour in which the research was undertaken.

It provides a protocol and a set of tools that allow the recognition of work-place competences that are highly valued by employers but not recognised formally. At the same time, it reflects back to the worker the depth of what they 'know' that enables them to do what they do and what they have the capacity to know, whether in pursuit of work-based training, change or career advancement. It helps practitioners to prepare students for work-placement so that they can maximise the value of the experience. It also helps curriculum developers to legitimise their course content and to underline its professional relevance or to uncover its irrelevance.

This is a timely intervention given the growing concern regarding the potential impact of automation across a broad spectrum of jobs. The history of industrial revolution demonstrates that the jobs made obsolete by new methods were replaced by more, previously unimagined jobs. This time it is different because of the emergence of artificial intelligence, capital automation, i.e. machines that can build machines, and the multiplier effect where a single person's intervention e.g. computer programmer, can be disseminated across countless millions of devices without any extra effort. The risk is that the consequent attrition in jobs will render many people unemployed and more importantly, unemployable in their own perception. The function of the Workplace Context-Complexity Protocol will enable individuals to highlight what they 'know' as distinct from what they 'do' to be recognised formally, and to identify the ways in which it can be re-contextualised to embrace new learning and to be re-purposed to meet the challenges of the emerging and exhilarating new World of the 21st Century.

APPENDIX 1

NATIONAL SURVEY OF PEOPLE AT WORK IN IRELAND

The first few questions are about which part of the workforce you represent. Please do not enter any personal information.	
1. What is your job Title (6 words max)?	
2. What is your gender? Please tick the box that most closely matches your answer	
Female	
Male	
3. Please indicate your age range in years	
Under 20	
21 – 40	
41 – 60	
Over 60	
4. What is your highest level of education (Professional or FETAC level equivalent)?	
None (Informal)	
Primary School	
Lower Secondary School	
Upper Secondary School	
Post leaving Certificate (PLC)	
Certificate	
Diploma	
Ordinary Degree	
Honours Degree	
Masters	
PhD	
Post Doctoral	
Other (Please specify)	

 | DOI:10.1163/9789004381766_007

5. What is the highest level at which you studied mathematics?	
None (Informal)	
Primary	
Lower Secondary	
Upper Secondary	
3rd Level	
Professional	
Other (Please specify)	

6. Please mention the skills that are key to your job.
i) ii) iii)
7. Thank you for answering these questions. This will help me to ensure that all aspects of the workforce are represented. The next section has to do with how people think about their jobs. There are no wrong answers. You are an expert in your job, so your first response is probably the best. This part should take about 5 minutes only, please continue.

8. How do you think about your work? The following are statements about how people think about their work. Please mark the answer in the box that best represents your opinion.						
	Don't know	Strongly Disagree	Disagree	Agree	Strongly Agree	Don't understand the question
Quantities and measurements are a good basis for comparing two things						
I pick out mathematical information to do my job						
The work I do is straight forward						

	Don't know	Strongly Disagree	Disagree	Agree	Strongly Agree	Don't understand the question
The ability to calculate and estimate is important in my job						
Diagrams, plans, models, charts and charts are useful when discussing work with colleagues						
Differences in shapes and patterns between jobs are a useful way of comparing them						

9. We are more than half way through, please continue.						
	Don't know	Strongly Disagree	Disagree	Agree	Strongly Agree	Don't understand the question
I use mathematical information in work						
Graphs, charts, maps and schedules contain information I need for my job						
The solution to a work problem is based on data and likelihood						
I follow procedure to do my job						
Maths plays an important part in my work						
It takes time to learn the lingo we use in work						

10. These are the last few items concerned with how you think about your work.						
	Don't know	Strongly Disagree	Disagree	Agree	Strongly Agree	Don't understand the question
Thinking about how a problem came about and how effective the solution was is good preparation for the next time a problem occurs						
There is a procedure for dealing with many of the situations that arise in work						
I adapt 'off the job' training I get for the job that I actually do						
I pass on to another person any work problems that involve quantities or numbers						
Knowing what to expect in work can give an early indication of a problem						
Work problems can be solved with the help of graphs, or timetables, or maps or numbers						
Mathematics is important in the World generally						

Thank you for your help so far. This is the last part of the survey. It should take only another 5 minutes or so. The following statements have to do with **how often** you do the things suggested by the following statements. Please tick the box that most closely matches your opinion. As before, there are no wrong answers because I am interested in your opinion about your job.

11. Please note: The answers to the following statements have to do with How Often .

	Never or Hardly ever	About once a year	About once a month	About once a week	About every day	Don't understand the question
I interpret information from a variety of sources to form an opinion about a work situation						
I use measurements to describe the solution to a work problem						

12. Please note: The answers to the following statements have to do with How Often.

	Never or Hardly ever	About once a year	About once a month	About once a week	About every day	Don't understand the question
I apply ready made solutions to familiar problems that arise in work						
I measure dimensions in work						
I notice a problem in work when the measurements and dimensions seem to be out of line with the usual						
From among possible solutions, I select the one most likely to solve the immediate problem						

	Never or Hardly ever	About once a year	About once a month	About once a week	About every day	Don't understand the question
When trying to solve a work problem, I take into account that a change in one part of the problem could have an affect on another part						
I use my knowledge of the system to trace fault back to the cause.						

13. We are more than half way through this section, only a few more minutes please.						
	Never or Hardly ever	About once a year	About once a month	About once a week	About every day	Don't understand the question
I think about the shapes involved in a problem to find the solution						
I notice there is a problem when some aspect of my work differs from what I would normally expect						
I use my knowledge of how different parts of the system affect each other to find a solution to a problem						
I convert work information from one measuring system to another						

14. Please note: The answers to the following statements have to do with How Often						
	Never or Hardly ever	About once a year	About once a month	About once a week	About every day	Don't understand the question
I work with amounts and quantities						
I discuss work drawings, schedules, or spreadsheets with colleagues to make sense of them						
I use the information I get from a variety of sources to make judgements about work						
I figure out the affect that one quantity may have on another in work						
I read charts, graphs, sketches and the like, to get information about my work						

15. Please note: The answers to the following statements have to do with How Often						
	Never or Hardly ever	About once a year	About once a month	About once a week	About every day	Don't understand the question
I detect problems in work when something doesn't seem to fit in with the usual pattern						
I get work information in the form of measurements and proportions						
I notice there is a problem in work when the shape of some part of it is different from usual						

	Never or Hardly ever	About once a year	About once a month	About once a week	About every day	Don't understand the question
I merge work data from different sources to create an overall report						
I solve a problem by exploring data for a likely explanation						
I make estimates about the information contained in charts, graphs, diagrams, timetables and the like						

16. Just 3 more sets to go, please continue						
	Never or Hardly ever	About once a year	About once a month	About once a week	About every day	Don't understand the question
I discuss quantities associated with my job with work colleagues						
I use reasoning skills on the numbers aspect of a problem to try to find a solution						
I find what I want by sorting information into sequence						
I use the shapes involved in a problem to try to find a solution						
I detect problems in work when the data does not seem to fit with normal expectations						
I describe a work situation using the data involved						

17. Almost finished.						
	Never or Hardly ever	About once a year	About once a month	About once a week	About every day	Don't understand the question
I use diagrams, sketches, models and the like when I communicate about work						
I get work information in the form of quantities and numbers						
I use reasoning skills on the shapes involved in a work problem when trying to figure out a solution						
I measure things						

18. This is the last set. Thank you very much for your patience and help.						
	Never or Hardly ever	About once a year	About once a month	About once a week	About every day	Don't understand the question
I describe some things in work using numbers and mathematical ideas						
I apply ready made solutions to familiar problems that arise in work						
I make work judgements around priority, urgency, importance and so on, on the basis of the numbers involved						
I use my knowledge of shapes when trying to find a solution to a work problem						

APPENDIX 1

	Never or Hardly ever	About once a year	About once a month	About once a week	About every day	Don't understand the question
I keep track of potential problems by comparing the actual with the expected outcomes						
I check the quantities aspect of a work solution after a solution has been implemented to see how well it has worked						

APPENDIX 2

BUSINESS ACTIVITY MODEL – EXEMPLAR

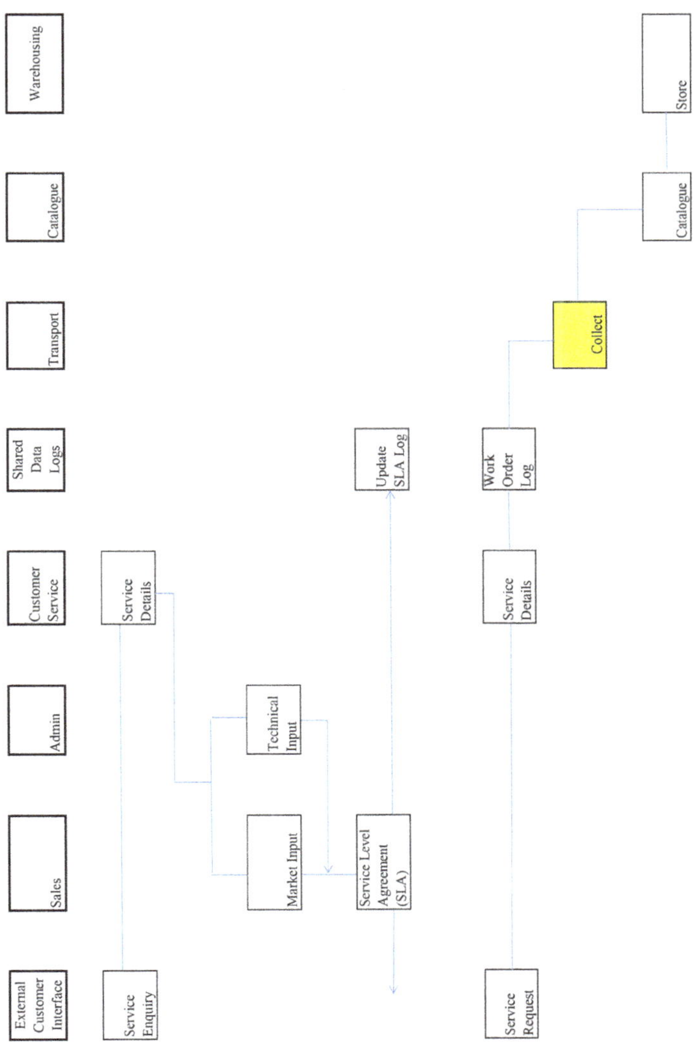

 | DOI:10.1163/9789004381766_008

APPENDIX 3

WORK PRACTICE MODEL – EXEMPLAR

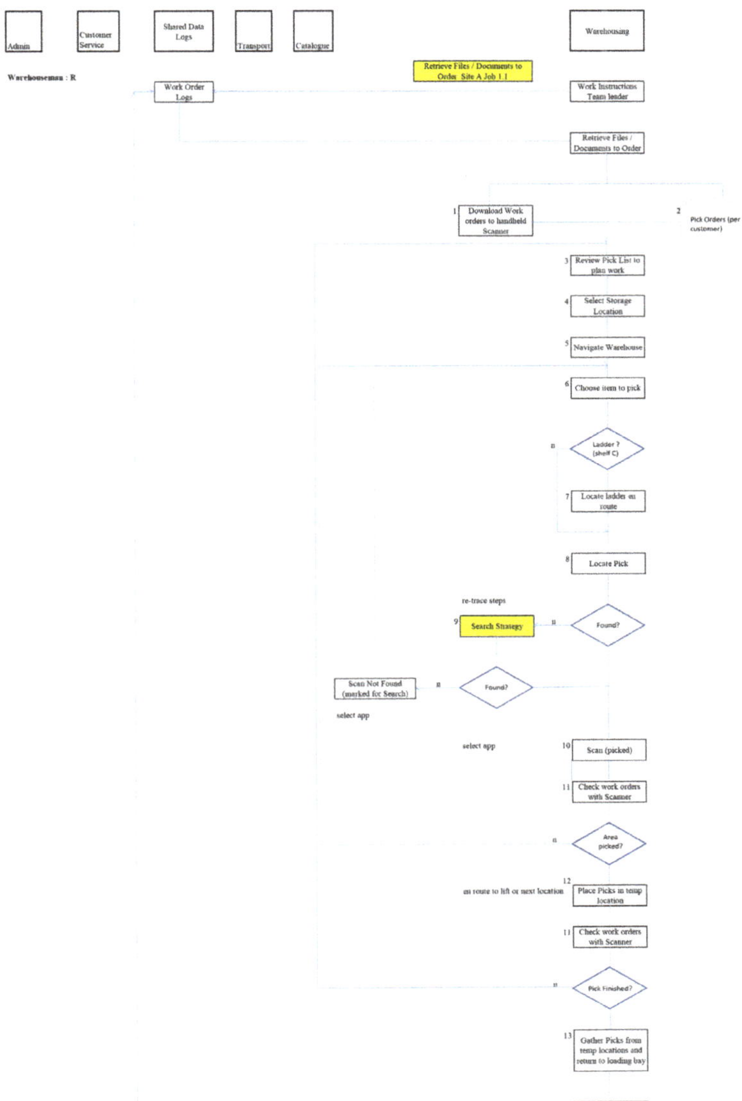

 | DOI:10.1163/9789004381766_009

APPENDIX 3

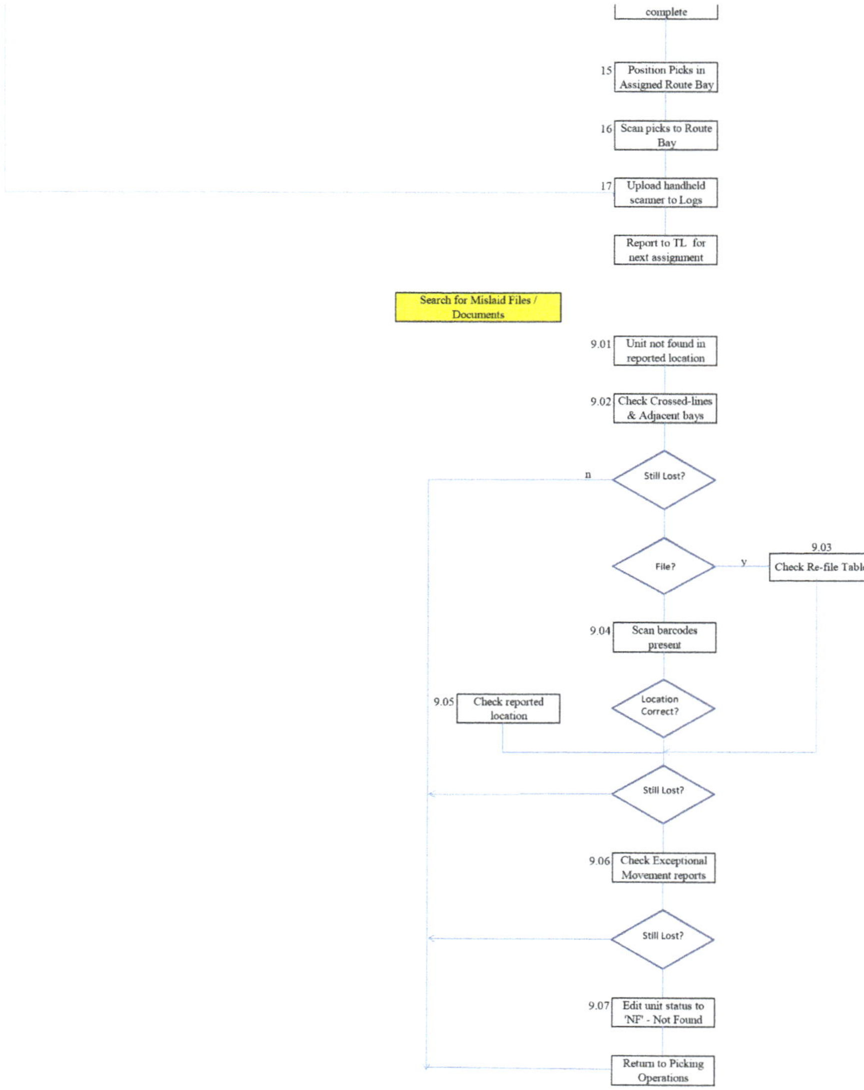
complete
15 Position Picks in Assigned Route Bay
16 Scan picks to Route Bay
17 Upload handheld scanner to Logs
Report to TL for next assignment
Search for Mislaid Files / Documents
9.01 Unit not found in reported location
9.02 Check Crossed-lines & Adjacent bays
n
Still Lost?
File?
y
9.03
Check Re-file Table
9.04 Scan barcodes present
9.05 Check reported location
Location Correct?
Still Lost?
9.06 Check Exceptional Movement reports
Still Lost?
9.07 Edit unit status to 'NF' - Not Found
Return to Picking Operations

APPENDIX 4

DETAILED TASK ANALYSIS

Observation Instrument : Detailed Task Analysis - Sequence (by sequence of tasks)

Site Code: A Job: 1.1Warehouse Picker

Job Description : 1.1 Retrieve Documents and Files

Instructed by work orders and guided by handheld scanner, retrieve files and documents that have been stored amongst 3 million other documents. Locations are identified by a 10 digit code, files by a further 8 digits, a third 8 digit code may be needed to identify the container of the file.

Boxes may also bear an owner provided identification number / barcode or text. Location codes are hierarchical by Building, Floor, Row, Bay and shelf. There are 3 buildings each with 3 or more floors, each floor has an access stairwell and is served by a centrally located goods lift. Each floor has up to 52 rows, each with 26 bays, each bay has a series of numbered columns of shelves on 3 leves, A,B and C at the top. Example LEV118J10C means that the building is LEV, on floor 1, row 18, bay J, column 10, top shelf. There are also Vaults and Data storage locations on some floors. Work orders are customer specific. There may be more than 1 work order for a customer. Customers are assigned to delivery routes. The scanner is a representation of the work orders that have been assigned to the picker. The information on the scanner is presented in the order of the location code. However, this does not inform the picker as to the most efficient way or picking, because it does not take into account the load at each location, its proximity to access and egress, the lift, openings in the rows (one in the middle and (sometimes) one at each end, nor does it take account of whether it is more efficient to complete all rows on one side before crossing to the other. When a file is not found, the picker will look for it as thoroughly as the time allows, giving rise to the NF category of CBA (couldn't be arsed)

Task no		Q&N	S&S	P&B	Dh&C	Other
1	Download Work Orders to Scanner in alphabetical sequence of the location code					
1.1	Access Scanner Centre Ops software			√		ICT
1.2	Select Download option			√		ICT
1.3	Enter User Id		√			ICT
1.4	Place in download Cradle (or use wireless if available), ensuring that contact points meet to complete circuit		√			ICT
1.5	Complete operation and remove from cradle		√	√		ICT
2	Collect hard copy work orders, (2 copies) printed by reference to Customer / Route from warehouse Team Leader					
2.1	Review Work Orders for planning information, making note of type of item, volume and distribution	√	√	√	√	
2.2	Collect a trolley if the volume of picks justifies it. This is a trade off between the carrying capacity of the trolley, the volume to be picked and the convenience of freedom of movement on stairs etc	√	√	√		
3	Review Scanner to plan Pick strategy					ICT
3.1	Scanner gives a list of the Picks in alphabetic order indicating the building, floor, row, bay and shelf. Richie synthesises the scanner information and the work order information to optimise the sequence of picks, taking into account the layout of the buildings / floors and the volume to be picked in each area	√	√		√	
4	Select Storage Location sequence					
4.1	Plan the route through the warehouse. The scanner is guided by the hierarchy of the storage locations, but this is not always the most efficient, because it does not account for nearest neighbour, weight or volume of the pick. Following the scanner would result in all the picks being accumulated, terminating at the furthest point away from the loading bay. The experienced picker, will vary the route to ensure that each area picked terminates at or near the stairwell or lift, depending on the next area chosen to be picked.	√	√	√	√	
5	navigate warehouse					
5.1	There are 3 warehouse buildings, each differing in layout, with a joint capacity of 3 million boxes. Plus fireproof vaults for media and high value documents, e.g. property deeds. Richie interprets an 8 digit code to navigate the warehouse. Typically there are 52 rows in each floor, 26 bays in each row, up to 20 sections in each row, containing shelving at 3 levels on which boxes are stacked 3 high and 2, 3, or 4 deep depending on the building	√	√	√		
5.2	Richie takes into account the layout of the building and organises his route so that it concludes near the stairwell or lift or access to the next building		√	√		
6	Choose item to Pick, set scanner to 'identify and confirm'					
6.1	Set Scanner to 'Identify and Confirm'			√		ICT
6.2	Select an item from the scanner			√	√	
6.3	Notice if it is located on the top shelf, level C		√	√		
6.4	Choose whether to switch on the light in preference to using the scanner light. It may be too cumbersome to switch the light off if Richie is carrying 2 or 3 boxes	√	√	√		
7	Locate Ladder when necessary					
7.1	Interpret location code to decide if ladder is required (Shelf level C)			√		
7.2	There are several ladders on each floor, but not stored in a customary location. The practice is to allow the ladder protrude from amongst the racking and be more easily visible		√			
8	Pick item - Box or File					
8.1	Set the scanner to 'Scan Picked'			√		
8.2	Boxes located on the top shelf may be stored amongst 26 others. Create a temporary shelf by pulling out boxes that are located on the middle shelf. Use this temporary shelf instead of placing boxes on the floor, to search through the bay area. Care must be taken if vacant spaces in adjacent bays are used during the retrieval process and this may lead to boxes being mislaid, hence search step 1.1.09.02	√	√	√		
8.3	If the item is a box, scan the barcode to register that it has been picked					ICT
8.4	If the item is a file, first locate the box in which it is filed, open the box and riffle through the files by reference to the number printed under the barcode. Remove the file from the box and scan it as picked	√		√	√	ICT
9	Initiate Search Strategy if item is not in reported location, following each step in turn until the missing item has been found or the options have been exhausted. The search strategy is informed by previous experience of the circumstances under which units have been found, having been missing according to the storage database					
	If time is pressing, go to 1.1.09.07 - 'mark unit as NF'					
9.1	Step 1 Confirm unit not found by ensuring that the correct barcode is being scanned and not one supplied by the unit owner or previous provider	√		√	√	ICT
9.11	If time is pressing, go to 1.1.09.07 - 'mark unit as NF'					
9.2	Step 2. check crossed-lines and adjacent bays. This is to see if the box has been incorrectly replaced by a previous picking activity	√		√		
9.21	If time is pressing, go to 1.1.09.07 - 'mark unit as NF'					
9.3	Step 3 - if missing Files, check re-file table for missing files					

 | DOI:10.1163/9789004381766_010

APPENDIX 4

Observation Instrument : Detailed Task Analysis - Sequence (by sequence of tasks)

Site Code: A Job: 1.1Warehouse Picker

Job Description : 1.1. Retrieve Documents and Files

instructed by work orders and guided by handheld scanner, retrieve files and documents that have been stored amongst 3 million other documents. Locations are identified by a 10 digit code, files by a further 8 digits, a third 8 digit code may be needed to identify the container of the file.

Boxes may also bear an owner provided identification number / barcode or text. Location codes are hierarchical by Building, Floor, Row, Bay and shelf. There are 3 buildings each with 3 or more floors, each floor has an access stairwell and is served by a centrally located goods lift. Each floor has up to 52 rows, each with 26 bays, each bay has a series of numbered columns of shelves on 3 levels, A,B and C at the top. Example LEV118J10C means that the building is LEV, on floor 1, row 18, bay J, column 10, top shelf. There are also Vaults and Data storage locations on some floors. Work orders are customer specific. There may be more than 1 work order for a customer. Customers are assigned to delivery routes. The scanner is a representation of the work orders that have been assigned to the picker. The information on the scanner is presented in the order of the location code. However, this does not inform the picker as to the most efficient way or picking, because it does not take into account the load at each location, its proximity to access and egress, the lift, openings in the rows (one in the middle and (sometimes) one at each end, nor does it take account of whether it is more efficient to complete all rows on one side before crossing to the other. When a file is not found, the picker will look for it as thoroughly as the time allows, giving rise to the NF category of CBA (couldn't be arsed)

Task no.		Q&N	S&S	P&R	Dh&C	Other
9.31	The re-file table contains files that have been returned by drivers but not yet returned to their original box. There may be hundreds of files in this category and in no particular order. The picker may be familiar enough with the customer to recognise the file by its outward appearance as belonging to the customer and in this way can narrow down the search.	√	√	√		
9.4	Step 4. Scan boxes in the bay to verify reported location , taking care to scan the host company's 'own' barcode.					ICT
9.41	Set scanner to 'identify and confirm'			√		ICT
9.42	Scan target item. The scanner will return a location for each box scanned				√	
	If time is pressing, go to 1.1.09.07 - 'mark unit as NF'					
1.1.09.050	Step 5. if the location reported does not confirm the actual location: continue the search in the reported location on the basis that it could have been swopped with the box that is 'missing'		√	√		
1.1.09.051	Check other barcodes for recognition by the storage database				√	ICT
	If time is pressing, go to 1.1.09.07 - 'mark unit as NF'					
9.6	Step 6 check exceptional movements report					
9.61	Exceptional movements reports a breakdown in the logic of sequence of movements reported to the system. For example if a unit , whose previous status was 'permed out' (returned to the customer) the next reported movement should be collected from the customer and not 'located is storage location ... Such exceptions have indicated in the past that an incorrect barcode was scanned, passing incorrect information to the system. In the case of a missing box, the wrong barcode could have been permed out and returned to a different location all the while the box's original barcode reports an unchanged original location. by checking the disposition of illogically returned boxes, the missing box may be found.			√	√	ICT
9.62	Go to the location of the exceptionally returned items and check them for correctness		√	√	√	
	If time is pressing, go to 1.1.09.07 - 'mark unit as NF'					
9.63	if found, correct box movement history where necessary by creating missing movements or rescanning correct barcode			√	√	ICT
9.7	Step 7 edit unit status to 'NF' - Not Found					
9.71	Missing items is a substantial problem for the document storage and management business. Re-searching for missing items beyond the steps above is a specialist activity, conducted at weekends and by people with more access to the storage database than the warehouse picker. By a process of elimination, they can isolate all of the storage movements that were performed on the same day as the missing box and by the same person, they can match exceptions and have other techniques that have a high level of success in locating missing items				√	
10	Scan status of item					
10.1	Set scanner to 'scan picked'			√		ICT
10.2	scan item and place it with other retrieved items.		√		√	ICT
11	Check scanner for next pick					ICT
11.1	Identify and Interpret next pick code to determine whether there is more to be had from the same box or nearby, and to determine in which direction the next pick lies.		√	√	√	ICT
11.2	close box and return to shelf		√			
11.3	replace any boxes used as temporary 'shelf'		√			
11.4	Return Ladder ensuring that it is visible at the end of the row.		√			
12	Place picks in strategic temporary location					
12.1	Planning the pick for each area takes into account the location of stairwell, lift and the access to adjacent building. Rather than accumulate all items as the pick continues, common practice is to position picked items in a temporary location that suits the pattern the picker has decided to work in. Each picker has his 'own' preferred temporary location and this is respected by his peers. This is not in the standard operating procedures.		√			
13	assemble picks in preparation for leaving area					
13.1	Gather items from all the temporary locations and locate them about the lift access.	√	√			
13.2	Starting at the top level. Count the number of pieces and load them into the goods lift stopping at each floor on the way to the loading bay.	√		√		
13.3	Unload the lift and bring the items into the loading bay area, keeping the items collected from each level separate (in case of discrepancy)		√	√		
14	Check work order completeness					
14.1	Sort scanner into Work order sequence			√	√	ICT
14.2	verify that each item has been picked for each work order (simple count)	√		√		
14.3	in the event of a work order being incomplete, the scanner will report whether an item has been picked indicating that some pieces may not have been assembled from the various temporary locations. Otherwise Initiate search routine if time permits				√	
15	Position Picks in assigned Route bay					
15.1	Resort the work orders into Route Code (there are 7 city routes and 1 country)	√		√		
15.2	Assemble the items by reference to the work order and place them in the delivery bay proper to that route	√	√		√	

Observation Instrument : Detailed Task Analysis - Sequence (by sequence of tasks)

Site Code: A Job: 1.1Warehouse Picker

Job Description : 1.1. Retrieve Documents and Files

instructed by work orders and guided by handheld scanner, retrieve files and documents that have been stored amongst 3 million other documents. Locations are identified by a 10 digit code files by a further 8 digits, a third 8 digit code may be needed to identify the container of the file.

Boxes may also bear an owner provided identification number / barcode or text. Location codes are hierarchical by Building, Floor, Row, Bay and shelf. There are 3 buildings each with 3 or more floors, each floor has an access stairwell and is served by a centrally located goods lift. Each floor has up to 52 rows, each with 26 bays, each bay has a series of numbered columns of shelves on 3 levels, A,B and C at the top. Example LEV118J10C means that the building is LEV, on floor 1, row 18, bay J, column 10, top shelf. There are also Vaults and Data storage locations on some floors. Work orders are customer specific. There may be more than 1 work order for a customer. Customers are assigned to delivery routes. The scanner is a representation of the work orders that have been assigned to the picker. The information on the scanner is presented in the order of the location code. However, this does not inform the picker as to the most efficient way or picking, because it does not take into account the load at each location, its proximity to access and egress, the lift, openings in the rows (one in the middle and (sometimes) one at each end, nor does it take account of whether it is more efficient to complete all rows on one side before crossing to the other. When a file is not found, the picker will look for it as thoroughly as the time allows, giving rise to the NF category of CBA (couldn't be arsed).

Task no.		O&N	S&S	P&K	Dh&C	Other
15.3	Place a copy of the completed work order with the items for delivery					
16	Change status of pick to 'scanned to loading bay'					
16.1	Reset scanner to confirm location 'group' function					ICT
16.2	Scan Loading Bay Location code					ICT
16.3	scan each item in turn, changing the location status					ICT
17	Update work order data logs					
17.1	Return scanner to cradle in order to upload data					
17.2	ensure TL computer is set to receive work log update					ICT
17.3	Select function on the scanner					ICT
17.4	Complete upload					ICT
17.5	report for next assignment					
		16	25	34	19	

WORKPLACE CONTEXT COMPLEXITY PROTOCOL

	Workplace Context Complexity Protocol: Mathematics Knowledge, Skills & Competence (MKSC)				
Complexity Level Indicators	**Range 1–2**	**Range 3–4**	**Range 5–6**	**Range 7–8**	**Range 9–10**
Accountability attributes	**Strand: Accountability** (for the selection, computation, accuracy & application of mathematics techniques)				
Audit Materiality (the exposure to financial and systems audits, and the material consequences of making an error in selection, computation or application)	none – error free	peripheral audit- some low impact error – low risk of error	participant audit – mainly low impact, moderate risk of error	central audit – mainly high impact, moderate risk of error	critical audit – mainly high impact, high risk of error
Decision making (autonomy & discretion regarding the role of MSC)	none	partial, within narrowly defined limits	partial, within moderately defined limits	partial within broadly defined limits	full, responsibility for decision making
Initiative (the extent to which numerate behaviour is proactive or reactive)	no scope for initiative	scope within narrowly defined limits	scope within moderately defined limits	scope within broadly defined limits	unlimited initiative responsibility
Concreteness (how specific and direct the application of MKSC)	fully concrete by recognition / recall	concrete in reference and comprehension	mainly concrete with some integrated knowledge	mainly concrete, some abstraction / theory	some theoretical concepts, abstract thinking, significant underpinning theory
Judgement (latitude for making assumptions & providing for margins)	none	partial within narrowly defined limits	partial within moderately defined limits	full within broadly defined limits	unlimited application of judgement responsibility
Planning (influence of parallel & interdependent MKSC activities)	none	partial within narrowly defined limits	partial within moderately defined limits	full within broadly defined limits	unlimited planning responsibility
Responsibility (immediacy and ownership of the mathematics-containing situation & outcome)	no responsibility	partial responsibility within narrowly defined limits	partial responsibility within moderately broad limits	responsible within broad, vaguely defined limits	complete responsibility

DOI:10.1163/9789004381766_011

Clarity attributes	**Strand: Clarity** (the precision of the definition of the mathematics-containing problem/situation and the context in which it arises)				
Distracters (presence & possible influence of peripheral & irrelevant factors)	No distracting elements	narrow range of explicit distracting information	moderately broad range of explicit distracters, narrow range of embedded distracters	broad range of mainly embedded distracters	broad range of plausible, embedded and misleading distracters
Priorities (conditioning factors and possible constraints)	simple, clear, single fixed goal	narrow range of fixed goals	moderately broad range of changing goals	broad range of interdependent, volatile goals	broad range of competing, incompatible and volatile goals
Reflectivity (making meaning of the situation to inform selection, computation, accuracy and application of an appropriate MKSC)	no reflection required	narrow potential for reflection	somewhat broader scope for reflective practice	moderate extent of reflective practice	extensive reflective practice
Information Sources (variety, directness, format, significance & tangibility of information sources from which relevant information can be interpreted, extracted, deduced or inferred)	single information source	small quantity of homogeneous information sources	moderate quantity of mainly homogeneous & familiar sources	moderate quantity of mainly heterogeneous, multiple familiar sources	large quantity of heterogeneous multiple familiar and unfamiliar sources
Vision (to what extent the performance of the job is perceived as being part of a whole)	none	partial, confined by the job boundaries	partial, confined by preceding and following jobs	moderately broad view of the organisation's activities	broad view of the organisation's activities and position in industry
Information Completeness (the extent to which all, or sufficient, information is available to enable a mathematics-based solution, or partial solution)	all, and only relevant information provided	most information provided, narrow range of inferred information	most information provided, moderate range of inferred or assumed information	broad range of inferred, assumed or deduced information supplemented by information provided	all information completely inferred, assumed, deduced or abduced

Familiarity attributes	**Strand: Familiarity** (with the range of situations in which a mathematics-based solution may apply)				
Specificity (directness of the connection between the situation and MKSC technique)	single specific, recognised, unvarying, familiar	mainly specific	mainly specific, but with some abstraction	specialised, theoretical, analogical, with variation	abstract, theoretical, unfamiliar, wide variance
Principal Activity (explicitness of the main purpose of the job underpinned by appropriate MKSC)	individual in a closely defined activity	co-operating with co-located, known individuals, activity defined within limits	varied and specific, creative and non-routine, loosely defined individual or group activity	wide variety of functions, including diagnostic/ creative, loosely defined activities	professional, self-directed, leading edge activities
Elements (expectation of mathematics containing elements arising in sequence)	single, fixed element	narrow range of elements arising in a fixed sequence	varied range of elements arising in sequences that are moderately determined by internal factors	wide variety of elements arising in sequences mainly determined by known internal factors, influenced by few, known, external factors	extensive range of elements arising in familiar and unfamiliar sequences determined by known and unknown external factors
Facets (the dimensionality of the mathematics containing situation)	a single, unchanging facet	narrow range of familiar facets	moderately broad range of mainly familiar facets	broad range of facets, mainly unfamiliar	extensive range of unfamiliar facets
Group (personnel dimension of the problem-space and the implications for negotiation of solutions)	single person, acting alone	small, familiar group	large familiar group, mainly co-located	large, mainly unfamiliar group, mainly co-located	large mainly unfamiliar group, mainly multiple locations
Routine (extent of habit-forming practices & degree of challenge)	habit formed, unchanging routine	habit-formed sequence of routines	partially conditioned, reactive routine	mainly conditioned reactive and proactive routine	mainly non-routine

Stressors attributes	Strand: Stressors (depth and range of sources of stress in which mathematics operations are performed)				
Constraints (presence and extent of considerations that temper the selection and application of MKSC)	clear, simple, fixed constraints	narrow range of constraints, clearly defined	broad range of constraints, mainly explicit and clearly defined, but incomplete	moderate range of constraints, mainly inferred from internal conditions, ad hoc	broad range of constraints, inferred from internal and external conditions, mutually interdependent and ad hoc
Pressure (factors that shape the environment in which mathematics is selected & applied)	no pressure	loosely defined, peer expectations	moderate yet explicit time/ accuracy/ volume / productivity/ peer pressure	high compliance, quality and outcome pressure	extreme pressure
Problem – potential range (potential breadth, depth and range of problems and response-scope)	none	few and familiar problems with known solutions	moderate range of familiar problems	mainly unfamiliar problems	multiple novel problems
Solutions (potential breadth, depth and range of solutions and related mathematics topic)	one response to all problems	few and known familiar responses to familiar problems	mainly conditioned responses to moderate range of mainly familiar problems	conditioned responses to mainly unfamiliar problems	novel responses to unfamiliar problems
Sources of Stress (the range and extent of exposure to internal & external sources of stress that may influence the application of mathematics)	none	partial exposure to few internal sources	moderate exposure to mainly internal sources of stress	broader exposure to moderate range of mainly external sources of stress	relentless exposure to multiple external and internal stressors
Structure (command & control, requirement to report, seek official approval, possible sanction for error – when clearly defined, eases stress, otherwise, increasingly stress-inducing)	tightly defined	defined within narrow limits	defined within moderately broad limits	loosely defined, within broad limits	undefined

Volatility attributes	**Strand: Volatility** (of the mathematics-containing situations in which the selection, computation, accuracy & application of mathematics techniques occur)				
Conditionality (range of conditions that may modify a mathematics-containing situation)	simple, straight forward, no scope for conditioned responses	narrow scope for conditioned responses	moderate scope for responses conditioned by mainly known internal factors	all responses conditioned by internal, & mainly predictable external factors	all responses conditioned by known and unknown external & unpredictable factors
Demands (range and extent of competing demands and their influence on the selection, computation, accuracy and application of mathematics)	single, fixed demand	narrow range of demands	moderately broad range of demands	broad range of demands	broad range of competing demands
Diversity (presence or otherwise of extenuating circumstances relating to common features of diversity, whether social, cultural, political, physical, intellectual, educational etc.)	none, completely homogeneous and familiar	narrow scope for diversity	moderately broad expectation of diversity	broad expectation of diversity	completely diverse
Predictability (extent to which the status of a variety of factors may influence the role of mathematics and the desired outcome may be predicted)	fully predictable	mainly predictable	moderately un-predictable	mainly unpredictable	totally unpredictable
Range (the extent of the spectrum across which affecters may be anticipated)	no range	narrow range	somewhat broader	moderate	broad
Risk (the certainty with which the desired outcomes can be expected and exposure to potentially extreme negative consequences for MKSC)	no risk – certain outcome	low risk – within narrowly defined limits	moderate risk – within moderately broad limits	high risk, within broadly defined limits	high unbounded risk, volatile & unpredictable conditions
Stability (the stability of the factors that may contribute to the description of, and response to, a mathematics containing situation and the extent to which that informs action)	completely stable	mainly stable	moderately unstable	mainly unstable	totally unstable

SELF-PERCEPTION OF NUMERATE BEHAVIOUR

The following statements have to do with **how often** you do the things suggested by the following statements. Please tick the box that most closely matches your opinion. There are no wrong answers because I am interested in your opinion about your job.

	Never or Hardly ever	About once a year	About once a month	About once a week	About every day	Don't understand the question
I interpret information from a variety of sources to form an opinion about a work/ hobby situation						
I use measurements to describe the solution to a work/hobby problem						
I apply ready made solutions to familiar problems that arise in work/hobby						
I notice a problem in work/hobby when the measurements and dimensions seem to be out of line with the usual						
From among possible solutions, I select the one most likely to solve the immediate problem						

 | DOI:10.1163/9789004381766_012

	Never or Hardly ever	About once a year	About once a month	About once a week	About every day	Don't understand the question
When trying to solve a work/hobby problem, I take into account that a change in one part of the problem could have an affect on another part						
I use my knowledge of the system to trace fault back to the cause.						
I think about the shapes involved in a problem to find the solution						
I notice there is a problem when some aspect of my work/ hobby differs from what I would normally expect						
I use my knowledge of how different parts of the system affect each other to find a solution to a problem						
I convert work/hobby information from one measuring system to another						
I work with amounts and quantities						
I discuss work/hobby drawings, schedules, or spreadsheets with colleagues to make sense of them						

	Never or Hardly ever	About once a year	About once a month	About once a week	About every day	Don't understand the question
I use the information I get from a variety of sources to make judgements about work/hobby						
I figure out the affect that one quantity may have on another in work/hobby						
I read charts, graphs, sketches and the like, to get information about my work/hobby						
I detect problems in work/hobby when something doesn't seem to fit in with the usual pattern						
I get work/hobby information in the form of measurements and proportions						
I notice there is a problem in work/ hobby when the shape of some part of it is different from usual						
I merge work/hobby data from different sources to create an overall report						
I solve a problem by exploring data for a likely explanation						
I make estimates about the information contained in charts, graphs, diagrams, timetables and the like						

	Never or Hardly ever	About once a year	About once a month	About once a week	About every day	Don't understand the question
I discuss quantities associated with my work/hobby with colleagues						
I use reasoning skills on the numbers aspect of a problem to try to find a solution						
I find what I want by sorting information into sequence						
I use the shapes involved in a problem to try to find a solution						
I detect problems in work/hobby when the data does not seem to fit with normal expectations						
I describe a work/ hobby situation using the data involved						
I use diagrams, sketches, models and the like when I communicate about work/hobby						
I get work/hobby information in the form of quantities and numbers						
I use reasoning skills on the shapes involved in a work/ hobby problem when trying to figure out a solution						
I measure things						

	Never or Hardly ever	About once a year	About once a month	About once a week	About every day	Don't understand the question
I describe some things in work/hobby using numbers and mathematical ideas						
I make work/hobby judgements around priority, urgency, importance and so on, on the basis of the numbers involved						
I use my knowledge of shapes when trying to find a solution to a work/hobby problem						
I keep track of potential problems by comparing the actual with the expected outcomes						
I check the quantities aspect of a work/hobby solution after a solution has been implemented to see how well it has worked						

APPENDIX 7

TASK CONTEXT TOOL

Task Context Tool: Tom's Job

Modelling the learner's Context – Worked example

Thinking about the big picture

Start: Tell me about where you work in *Documents'RUs*

What happens there? *We store documents and things for people, until they want them returned.*

Why (do people store their documents with you)? *They save on office space and rent we provide them with safe storage.*

What happens when they want they documents back? *They phone the office and tell them what they want, the office tells me and the warehouseman. He gets the documents or whatever, and I deliver them.*

How does he know what to get? *He gets a work order, same as me.*

Are they exactly the same*? No, his has a code to tell him where to get them, mine is the customers address and what I have to deliver.*

Is that all? *Well there would be a few work orders in the morning and a few more in the afternoon.*

Just a few*? Well it could be anywhere from 10 to 20, but usually around 15, 20 would be a very busy morning. You have to do all of them, but some customers are more demanding than others.*

How does the customer send the documents back*? Same thing, they phone the office and order a collection. That's on a work order too.*

What happens next? *The warehouse man leaves the items for delivery in my bay, so I check it and load the van and off I go to visit each customer on my list.*

What happens next? *When I come back I check in all the collections in the returns bay for refiling, and then I go to the canteen.*

Is that all? *No, there is usually another run in the afternoon and sometimes a few special collections which get special treatment.*

 | DOI:10.1163/9789004381766_013

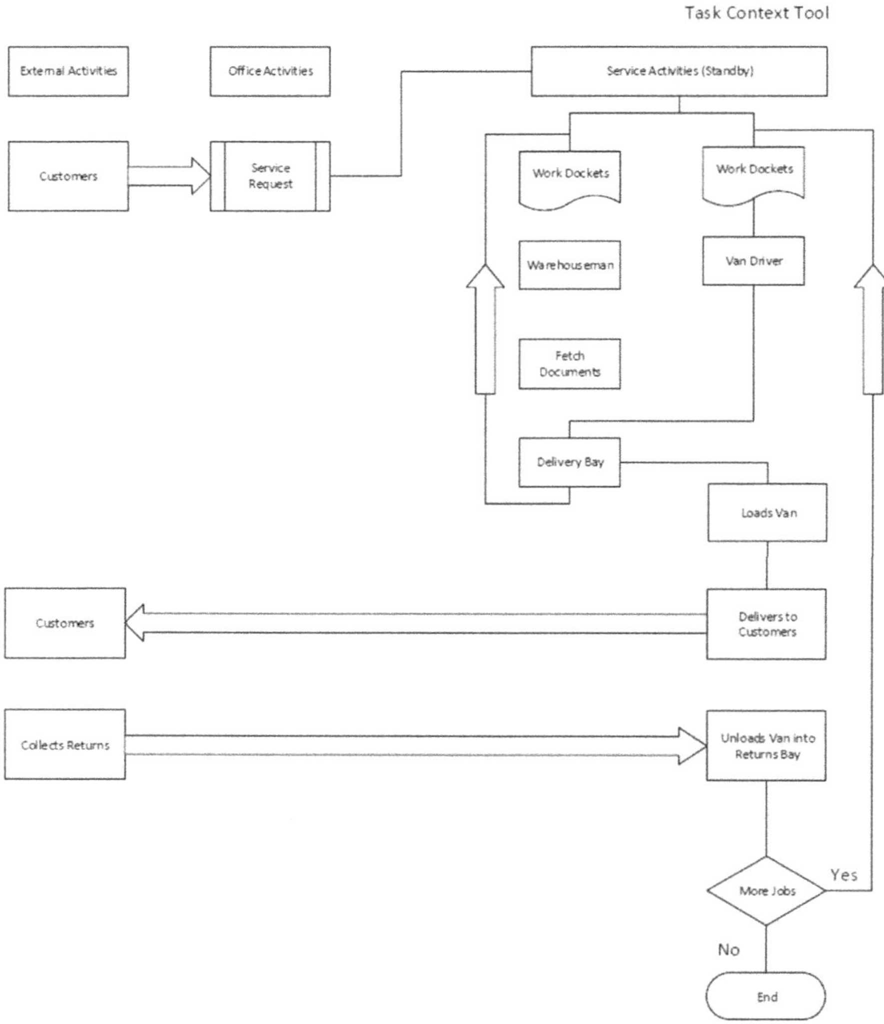

OK, let me check if I have this right so far

A person calls the company and asks for the service. You are given a work order that tells you what to deliver and to whom, so you go to the premises and deliver the service. You collect documents and computer backup media and go back to the office, and check them in to the returns bay, and wait for the next job – is that right so far?

There are obvious holes in the narrative, but it is not necessary to get every detail at this point. It is clear that the worker gets his instructions from someone else, he carries them out, accounts for them and waits for the next job. There are hints that he has expectations of the sequence of events and their interdependence, and that his actions may be conditioned or prioritised in some ways. It is likely that he may have more than one job on the same day, and maybe of different kinds, but that level of detail is for the Task Exploration Tool.

APPENDIX 8

TASK EXPLORATION TOOL

Task Exploration Tool (Tet): Worked Example

This about the task detail

OK so you arrive for work in the morning – what's the first that happens? *I collect my work orders from my pigeonhole and look them over.*

Why? *I want to see how my first shift is going to work out.*

What do you mean? *Well, if I notice that I have a delivery near someone else's route, I might be able to get them to do it for me, then I have to figure out the best way to do what I have left.*

Is it the same every day? *There are regular customers which I see a lot, and others not so often – you can't really tell until you see the work orders.*

What happens next? *I shuffle the orders around into the sequence that I think is easiest for me.*

How do you decide that? *I know where they are in the city and I just make it handy for myself.*

What do you take into account? *There's a few things like traffic, parking restrictions, pedestrian ways, all the roadworks on the LUAS (new public transit system), whether the customer will be ready with their backup tape – there's no point in me calling in if they are not ready – that sort of thing.*

Is there anything else? *There are a few customers who make late orders from time to time, so I check if there are any lates, because I am better off knowing before I leave the depot than when I am on the road.*

How often does that happen? *Oh, every other day, but it only takes a minute or two to check and it saves me a lot of hassle if I know what I have to do in advance.*

Why is it important to figure out how you are going to do the work orders? *That's how I know how to load the van – last in is first out.*

OK, so you have sorted your work orders, what happens then? *I go to the canteen and have a cup of tea and some toast – because I won't be back in time for break. Besides this is where I meet the other drivers and maybe do some swaps.*

 | DOI:10.1163/9789004381766_014

What happens next? *I download my work orders onto the handheld scanner.*

Why? *I have to 'beep' every barcode that I have to deliver or collect, so that tells the computer about the deliveries and collections that I am responsible for. Every time an item moves in or out or around the warehouse, it has to be beeped so that the system knows where it is.*

What happens next? *I go to my loading bay and check that every item on the work orders is ready for me.*

How do you do that? *I count the number of items on the work orders and then I count the items in the items in the bay to see if they are the same.*

What happens if the numbers are not the same? *They hardly ever are, because we don't leave important stuff out overnight.*

What do you mean? *Computer backup tapes, legal files, property deeds and so on, are kept in the vaults so I have to get them myself and take special care of them – it's a security thing.*

Where are the vaults? *There's a few of them, they are for different kinds of things.*

Why? *They get extra protection, because they are very expensive to replace.*

How much is expensive? *It could be half a year's wages – so we have to be very careful with them.*

How do you take care of them? *Well, each item has its own barcode, each location has its own barcode too, that includes the particular vault, the loading bay, waiting to be loaded, loaded on the van and finally, delivered to the customer. Each time the item is moved it has to be scanned to the new location – so that if anything goes wrong, we can track its history.*

So will all the files and boxes going in and out of the van, how do you find the box you need from among all the other boxes in the van? *Well I have an idea that it should be near the last delivery I made, because that's the way I loaded the van, but some customers use their own boxes, or in the case of computer backup media, they have different shapes cases and drums. As well as that, some customers boxes are heavier than others, so I have to get the trolley out, otherwise I could hurt my back and I don't get paid if I am not in work.*

Do things go wrong? *Sure, you just have to make sure it's not your fault. Imagine if something is put in the wrong place or some movement wasn't scanned, the trouble that would cause.*

Can you give me an example? *Say I scanned something on to the van, and then didn't load it, I would only realise that when got to the customer and I would be faced with the choice of telling them I had made a mistake and did not bring everything they ordered, or go back to the warehouse and get it.*

How do you decide what to do? *It depends on the customer. Some would give you a hard time, others would understand. Maybe they would be less likely to understand a mistake if it had happened before – and this is a very competitive business, customers are being tempted by the opposition everyday.*

How would that be sorted? *First I have to make sure that I actually did leave it in the warehouse – I could have delivered it to someone else – that would be a disaster. Then I know what my choices are for example, I could try to charm the customer, or get someone else to deliver or return to the depot and get it myself – it depends.*

You mentioned collections, how does that work? *The customer phones in to order that files are to be collected –but you never really know what to expect.*

What do you mean? *Some customers will say they have 10 folders for collection, but that could turn out to be 25 or 30 folders.*

How does that happen? *It varies. For example, the customer could change their mind, but what often happens is that one department notices that files are being lined up for collection, and they just add their returns to it.*

How is that a problem? *Well you get to know which customers do that and you make allowances for it. Strictly speaking, I should only collect what's on the work order, but I have to keep the customer happy too.*

What happens next? *I beep the files for collection to record the fact that I have collected them, print a receipt for the customer and one for the work order, and load them on the van.*

What if they are new files, not stored by your company before? *Then I issue new bar codes, stick them on to the files, and beep them as new, to be catalogued, that I have collected them. When I get back to the warehouse I have to put them in a special holding bay while they wait for the girls in the office to catalogue them.*

What do you do after you have done all the collections and deliveries? *I go back to the warehouse and unload the van. I have to make sure to beep all the items to record that I am finished with them – otherwise, the next time someone is looking for a file the computer would tell them I still have it in the van.*

Is there anything else? *Sometimes I might get a call from the office to do a collection that was not on the work order sheet. That always means it is an important customer so I have to make whatever changes are necessary to my plan to fit that in.*

What's the last thing you do? Oh I connect my handheld scanner to the computer and upload all the details to the computer.

Ok so have got this right? – then review the model with the student and make corrections clarifications as necessary.

APPENDIX 8

Task Exploration Tool: What The Learner Actually Does

Process

This part of the process is about taking a closer look at what the learner/worker actually does, in order to uncover the mathematics ideas that s/he relies on. This is more individual and is seeking a fairly detailed account of the processes and procedures involved, rather than a list of the names of duties. It is useful to position yourself as a trainee being mentored by the learner to do the job. Listen out for codified references, jargon and acronyms as they usually conceal a wealth of information. Most jobs have repetitive components, but the choice between procedures may be conditioned by factors that have to be evaluated, e.g., time, distance, pressure, importance, priority, value, likelihood, risk and so on.

Pool of Question Types

- So you get a work docket – what happens next?
- How is that?
- Do you get only one docket at a time?
- What happens next?
- What do you mean by materials?
- How do you know what materials you need?
- How do you know you have everything (all the information) you need?
- Why?
- How (& why) might the office in contact with you while you are on the road?
- What happens when you get back to the office?
- Is your work straight forward?
- What happens if there is a mistake?
- What kinds of things go wrong?
- How do you avoid things going wrong?
- How do you choose between options?
- How do you make decisions?
- What thinks to you have to consider when making a decision?
- What are the risks of making a mistake?
- What sorts of mistakes are made?
- Is that the official way of doing things?

Dos and Don'ts

Do	Don't
Try and put yourself in the shoes of a person being trained in or mentored by the learner	Make assumptions about the individual's workplace or role
Think about detail	Offer possible answers to your own questions
Actively listen to what the learner is saying	Interrupt the answer
Pause at the end of the answer to give the learner a chance to add more detail	Anticipate the end of the answer
Ask for clarity	Accept 'dead–end' answers e.g. 'same routine everyday'.
Remember that very few work or hobby actions occur in isolation of other people or actions	Ask 'closed' questions
Model the Work practice	Ask 'leading' questions
Check your model with the learner	Give a hint that there is a 'right' answer
Check for completeness	
Check for coherence	
Check for unusual occurrences	
Remember, most people are proud of what they do and how well they do it, give them a chance to 'show-off' their clever innovations	

APPENDIX 9

LINKING TOOL

Link the numeracy underpinning the work component to the relevant mathematics domain and QQI level/SLO

Process

Notice in the AMT that each act described by the learner is numbered and contains a brief description. This should be detailed enough for you to have identified the underpinning mathematics domain. This is where your expertise and objectives for the learner come to the fore. By taking the processes and procedures described in the WPMT, and their underpinning mathematics into account, it will be possible to detect which mathematics domains are involved, and, depending on each case, the FETAC level and approximate SLO code to which it refers.

Work/Hobby Component	**Details**	**Mathematics Domain**				**QQI Level & SLO Code**			
		QN	**SS**	**PR**	**DC**	**SLO**	**SLO**	**SLO**	**SLO**
1.									
2.									
3.									
4.									
5.									
6.									
7.									
8.									
9.									
10.									
11.									
12.									
13.									
14.									
15.									

 | DOI:10.1163/9789004381766_015

REFERENCES

Benedicty, A., & Donahoe, S. (1997). Math curse revisited. In R. Richart, D. P. Wolf, J. Craven, & D. Balick (Eds.), *Through mathematical eyes: Exploring functional relationships in math and science* (pp. 71–101). Portsmouth, NH: Heinemann.

Benn, R., & Burton, R. (1993). Access mathematics: A bridge over troubled waters. *Journal of Access Studies, 8*, 180–190.

Bernstein, B. (1999). Vertical and horizontal discourse: An essay. *British Journal of Sociology of Education, 20*(2), 157–173.

Bourdieu, P. (1977). *Outline of a theory of practice* (23rd ed.). Cambridge: The Press Syndicate of the University of Cambridge.

Bynner, J., & Parsons, S. (1997). *Does numeracy matter?* Retrieved from http://www.eric.ed.gov/ERICDocs/data/ericdocs2sql/content_storage_01/0000019b/80/16/7a/21.pdf

Bynner, J., & Parsons, S. (2005). *Does numeracy matter more?* Retrieved from http://www.nrdc.org.uk/doc_2979.pdf & http://www.lifelonglearning.co.uk/mosergroup/annex.pdf

Cedefop. (2015). *European guidelines for validating non-formal and informal learning.* Retrieved from http://dx.doi.org/10.2801/008370

Central_Statistics_Office. (2013). *PIAAC 2012 survey results for Ireland from the OECD's Programme for the International Assessment of Adult Competencies.* Dublin: The Stationery Office. Retrieved from http://www.education.ie/en/Publications/Education-Reports/Programme-for-the-International-Assessment-of-Adult-Competencies-PIAAC-2012-Survey-Results-for-Ireland.pdf

Chipman, S., & Wilson, D. (1985). Understanding mathematics course enrolment and achievement: A synthesis of the research. In S. Chipman, L. Brush, & D. Wilson (Eds.), *Women and mathematics: Balancing the equation* (pp. 275–328). Hilldale, NJ: Lawrence Erlbaum Associates.

Coben, D. (2000). Mathematics or common sense? Researching 'invisible' mathematics through adults' mathematics life histories. In D. Coben, J. O'Donoghue, & G. E. FitzSimons (Eds.), *Perspectives on adults learning mathematics: Research and practice* (Vol. 21, pp. 53–66). Dordrecht: Kluwer Academic Publishers.

Coben, D. (2002). Metaphors for an educative politics: Common sense, good sense and educating adults. In C. Borg, J. Buttigieg, & P. Mayo (Eds.), *Gramsci and education* (pp. 263–290). Lanham, MD: Rowman & Littlefield Publishers.

Coben, D. (2005). *Adult numeracy: Shifting the focus a report and recommendations on adult numeracy in Scotland.* Retrieved from http://www.communitiesscotland.gov.uk/stellent/groups/public/documents/webpages/alo_shiftingthefocusreport.pdf

Coben, D., & Thumpston, G. (1995). Common sense, good sense and invisible mathematics. In T. Kjaergard & A. L. N. Kvame (Eds.), *The third international conference on the Political Dimensions of Mathematics Education – PDME III* (pp. 284–297). Norway: Caspar F A/Sorlag.

Cockcroft, W. H. (1982). *The Cockroft report: Mathematics counts.* Retrieved from http://www.dg.dial.pipex.com/documents/docs1/cockcroft02.shtml & http://www.dg.dial.pipex.com/documents/docs1/cockcroft01.shtml

Colleran, N., & O'Donoghue, J. (2007). Adult mathematics education and commonsense. *Adults Learning Mathematics an International Journal, 2*(2), 20–32.

Colleran, N., O'Donoghue, J., & Murphy, E. (2003). *Adult problem solving and common sense: New insights.* Paper presented at the Learning Mathematics to Live and Work in our World: Proceedings of the 10th International Conference on Adults Learning Nathematics, Ströbl, Austria.

D'Ambrosio, U. (1985). Ethnomathematics and its place in the history and pedagogy of mathematics. *For the Learning of Mathematics, 5*(1), 44–48.

Engeström, Y. (1987). *Learning by expanding: An activity-theoretical approach to developmental research.* Helsinki: Orienta-Konsultit.

Engeström, Y. (2001). Expansive learning at work: Toward an activity theoretical reconceptualization. *Journal of Education and Work, 14*(1), 1–24. doi:10.1080'13639080020028747

Engeström, Y., & Cole, M. (1997). Situated cognition in search of an agenda. In D. Kirshner & J. A. Whitson (Eds.), *Situated cognition: Social, semiotic and psychological perspectives* (pp. 301–310). Mahwah, NJ: Lawrence Erlbaum Associates.

Engeström, Y., Miettinin, R., & Punamäki, R. L. (Eds.). (1999). *Perspectives on activity theory*. Cambridge: The Press Syndicate of the Universitry of Cambridge.

Ernest, P. (1995, July 7). *Images of mathematics, values and gender: A philisophical perspective*. Paper presented at the Mathematics with a Human Face, University of Exeter, Exeter, United Kingdom.

Evans, J. (1999). Building bridges: Reflections on the problem of transfer of learning in mathematics. *Educational Stuidies in Mathematics: Special Issue on the Contexts of Mathematical Teaching and Learning, 39*(1–3), 23–44. doi:10.1023/A:1003755611058

Evans, J. (2000a). Adult mathematics in everyday life. In D. Coben, J. O'Donoghue, & G. E. Fitzsimons (Eds.), *Perspectices on adults learning mathematics* (pp. 289–305). Dordrecht: Kluwer Academic Publishers.

Evans, J. (2000b). *Adults' mathematical thinking and emotions: A study of numerate practices*. London: RoutledgeFalmer.

FÁS. (2009). *National skills bulletin: The expert group on future skills needs*. Retrieved from http://www.skillsireland.ie/media/egfsn090703_national_skills_bulletin.pdf

Fitzsimons, G. (2002). Technoloy, mathematics, and industry. In A. J. Bishop (Ed.), *What counts as mathematics? Technologies f ower in adult and vocational education* (Vol. 28, pp. 59–97). Dordrecht: Kluwer Academic Publishers.

Fitzsimons, G. (2005, January 6). Can adult numeracy be taught? A bernsteinian analysis. In M. Goos, C. Kanes, & R. Brown (Eds.), *Mathematics education and society: Proceedings of the 4th International Mathematics Education and Society Conference* (pp. 155–165). Brisbane: Griffith University, Centre for Learning Research.

Fitzsimons, G., Jungwirth, H., Maasz, J., & Schloeglemann, W. (1997). Adults and mathematics (Adult Numeracy). In C. Laborde (Ed.), *International handbook of mathematics education, part two* (Vol. 2, pp. 755–784). Dordrecht: Kluwer Academic Publishers.

Gates, D., & Bremicker, M. (2017). *Beyond the hype: Separating ambition from realiy in i4.0, 24*. Retrieved from https://assets.kpmg.com/content/dam/kpmg/xx/pdf/2017/05/beyond-the-hype-separating-ambition-from-reality-in-i4.0.pdf

Geertz, C. (1993). *Local knowledge: Further essays in interpretive anthropology*. London: Fontana Press.

Gramsci, A. (1971). *Selections from the prison notebooks* (Q. Hoare & G. Nowell-Smith, Eds.). London: Lawrence & Wishart.

Harris, M. (1991). Mathematics and the traditional work of women. In I. Gal (Ed.), *Adult numeracy development: Theory, research, practice* (pp. 269–302). Cresskill, NJ: Hampton Press.

Hoyles, C. (2008). Empowerment through mathematics in the workplace. *Reflect: The Magazine of the NRDC*, 20–21.

Hoyles, C., Noss, R., Kent, P., & Bakker. (2007). *Techno-mathematical literacies in the workplace*. Retrieved from http://www.tlrp.org/pub/research.html

Hoyles, C., Wolf, A., Molyneux-Hodgson, S., & Kent, P. (2002). *Mathematical skills in the workplace*. London: Science, Technology and Mathematics Council.

Hudson, C. (2008). *Working mathematics for the 21st century? A policy discussion paper*. Retrieved from http://www.ncetm.org.uk/Default.aspx?page=13&module=res&mode=100&resid=8570 & http://www.ncetm.org.uk/files/254447/Working+mathematics+for+the+21st+Century.pdf

Keogh, J. (2013). *Looking at the workplace through mathematical eyes* (Ph.D. thesis). Institute of Technology, Tallaght, Dublin.

Keogh, J., Maguire, T., & O'Donnoghue, J. (2015, May 31–April 1). *Graduate work-readiness in the 21st century*. Paper presented at the Higher Education in Transformation Conference, Dublin, Ireland.

Keogh, J., Maguire, T., & O'Donoghue, J. (2016). *Mathematics invisibility: A subject-centric cultural historical activity theory perspective* (in press).

Keogh, J., Maguire, T., & O'Donoghue, J. (2016). *Recontextualising mathematics for the workplace*. Paper presented at the International Congress of Mathematics Education, ICME 13, Hamburg, Germany.

Klinger, C. (2008). *Passing it on: Linking adult innumeracy to mathematics attitudes, low self-efficacy beliefs, and math-anxiety in student primary teachers.* Paper presented at the 15th International Conference of Adults Learning Mathematics, Philadelphia, PA.

Lave, J., & Wenger, E. (1991). *Situated learning: Legitimate peripheral participation.* Cambridge: Cambridge University Press.

Leont'ev, A. N. (1978). *Activity, consciousness, and personality* (M. J. Hall, Trans.). Englewood Cliffs, NJ: Prentice-Hall.

Maguire, T. (2003). *Engendering numeracy in adults mathematics education with a focus on tutors: A grounded approach* (Unpublished Ph.D. thesis). University of Limerick, Limerick, Ireland.

Maguire, T. (2004). *Numeracy and sport: Integrating numeracy and sport guidelines.* Retrieved from http://www.sttc.ie/publications/Numeracy%20and%20Sport1.pdf

Maguire, T. (2012). *Developing maths eyes: A successful model for building confidence in mathematics in a community.* Paper presented at the 2012 Adults Learning mathematics – A Research Forum 19th International Conference (ALM19). Synergy – working together to achieve more than the sum of the parts. Te Piringa – Má pango, má whero, ka oti, Auckland, New Zealand.

Maguire, T., & O'Donoghue, J. (2003). *Numeracy concept sophistication: An organising framework, a useful thinking tool.* Paper presented at the 10th International Conference on Adults Learning Mathematics: A Research Forum, Strobl, Austria.

Marr, B., & Hagston, J. (2007). *Thinking beyond numbers: Learning numeracy for the future workplace.* Adelaide: NCVEDR. Retrieved January 8, 2008, from http://www.ncver.edu.au/publications/1795.html

Marsick, V. J. (1997). *Current thinking on the learning organisation.* Paper presented at the 5th Annual International Conference on Post-Compulsory Education and Training, Brisbane, Australia.

McLeod, D. B. (1992). Research on affect on mathematics education: A reconceptualisation. In D. A. Grouws (Ed.), *Handbook of research on mathematics teaching and learning* (pp. 575–596). New York, NY: Mcmillan.

NALA. (2004). *Meeting the numeracy challenge: Report.* Retrieved from http://www.nala.ienumeracy_reportNALA2004.pdf

Niss, M. (1994). Mathematics in society. In R. Biehler, R. W. Scholz, R. Strasser, & B. Winkelman (Eds.), *Didactics of mathematics as a scientific discipline* (pp. 367–378). Dordrecht: Kluwer Academic Publishers.

NQAI. (2009). *Referencing of the Irish National Framework of Qualifications (NFQ) to the European Qualifications Framework for lifelong learning (EQF).* Dublin: National Qualifications Authority of Ireland. Retrieved from https://www.qqi.ie/Documents/Referencing%20Irish%20NFQ%20to%20the%20European%20QF%20for%20Lifelong%20Learning.pdf

O'Donoghue, J. (2000). Assessing numeracy. In D. Coben, J. O'Donoghue, & G. E. Fitzsimons (Eds.), *Perspectives on adults learning mathematics* (Vol. 21, pp. 271–287). Dordrecht: Kluwer Academic Publishers.

OECD. (2000). *IALS: Literacy in the information age.* Retrieved from http://www.statcan.gc ca/pub/89-588-x/4152886-eng.htm

OECD. (2004). *PISA 2003: Learning for tomorrow's world.* Retrieved from http://www.oecd.org/dataoecd/44/53/33691596.pdf

OECD. (2005). *ALL survey learning a living.* Retrieved from http://www.oecd.org/dataoecd/44/7/34867438.pdf

O'Sullivan, C., Robinson, P., Keogh, J., & O'Neill, J. (2016). *Models of re-engaging adult learners with mathematics.* Paper presented at the 18th SEFI Working Group Mathematics Working Group Seminar, Gothenburg, Sweden.

PIAAC Numeracy Expert Group. (2009). *PIAAC numeracy: A conceptual framework.* Retrieved from http://www.oecd.org/edu/workingpapers

QQI. (2012). *National framework of qualifications.* Retrieved from http://www.nfq.ie/nfq/en/

Roth, W.-M., & Lee, Y.-J. (2007). "Vygotsky's neglected legacy": Cultural-historical activity theory. *Review of Educational Research, 77*, 186–231. doi:10.3102/0034654306298273

Rychen, D. S., & Salganik, L. H. (2003). *Highlights from the OECD project Definition and Selection Competencies: Theoretical and conceptual foundations (DeSeCo)*. Retrieved from http://www.eric.ed.gov/ERICDocs/data/ericdocs2sql/content_storage_01/0000019b/80/1b/0a/37.pdf

Skovsmose, O. (1996). *Meaning in mathematics education* (4). Retrieved from

Smith, A. (2004). *Making mathematics count* (Final report). Retrieved from http://www.mathsinquiry.org.uk/report/MathsInquiryFinalReport.pdf

Strasser, R. (2003). *Mathematics at work: Adults and artefacts.* Paper presented at the Learning Mathematics to Live and Work in our World: Proceedings of the 10th International Conference on Adults Learning Nathematics, Ströbl, Austria.

van Groensteijn, M. (2002). *A gateway to numeracy: A study of numeracy in adult basic education.* Utrecht: CD-Beta Press.

Wake, G. (2005). *Functional mathematics: More than "back to basics"*. Retrieved from http://www.nuffield14-19review.org.uk/files/documents101-1.pdf

Wake, G., & Williams, J. (2007). Metaphors and models in translation between college and workplace mathematics. *Educational Studies in Mathematics, 64*(3), 345–371. doi:10.1007/s10649-006-9040-6

Wedege, T. (1999). To know, or not to know mathematics, that is a question of context. *Educational Studies in Mathematics, 39*, 205–277.

Wedege, T., & Evans, J. (2006). Adults' resistence to learning school versus adults' competences in work: The case of mathematics. *Adults Learning Mathematics an International Journal, 1*(2), 28–43.

Wenger, E. (2006). *Communities of practice: A brief introduction*. Retrieved from http://www.ewenger.com/theory/index.htm and http://www.ewenger.com/theory/index.htm

Williams, J., Clemens, S., Oleinikova, K., & Tarvin, K. (2003). *The skills for life survey: A national needs and impact survey of literacy, numeracy and ICT skills*. Norwich: The Stationery Office.

Williams, J., & Wake, G. (2007). Black boxes in workplace mathematics. *Educational Studies in Mathematics, 64*(3), 317–343.

www.ingramcontent.com/pod-product-compliance
Lightning Source LLC
Chambersburg PA
CBHW071447230926
51846CB00001B/375
9789004381742